APPLICATIONS OF GEOGRAPHIC INFORMATION SYSTEMS (GIS) FOR HIGHWAY TRAFFIC NOISE ANALYSIS

Case Studies of Select Transportation Agencies

November 2012

Prepared for:
Office of Planning
Federal Highway Administration
U.S. Department of Transportation

Prepared by:
Organizational Performance Division
John A. Volpe National Transportation Systems Center
Research and Innovative Technology Administration
U.S. Department of Transportation

ACKNOWLEDGMENTS

The U.S. Department of Transportation John A. Volpe National Transportation Systems Center (Volpe Center) in Cambridge, Massachusetts, prepared this report for the Federal Highway Administration's (FHWA) Office of Planning. The project team included Alisa Fine of the Volpe Center's Organizational Performance Division and Jaimye Bartak of Cambridge Systematics.

The Volpe Center project team wishes to thank the staff members from several organizations nationwide, each listed in Appendix A, for providing their experiences, insights, and editorial review. The time they kindly provided was vital in preparing the case studies and drafting this final report.

TABLE OF CONTENTS

ACKNOWLEDGMENTS 1

EXECUTIVE SUMMARY 4

INTRODUCTION 5
   Purpose 5
   Methodology 5
   Background 6

OBSERVATIONS 8
   State of the Practice 8
   Benefits 9
   Challenges 10
   Lessons Learned 12

CASE STUDIES 15
   California Department of Transportation (Caltrans) 15
   Florida Department of Transportation (FDOT) 18
   Maryland State Highway Administration (MDSHA) 22
   North Carolina Department of Transportation (NCDOT) 25
   Ohio Department of Transportation (ODOT) 25
   Tennessee Department of Transportation (TDOT) 34
   Virginia Department of Transportation (VDOT) 34
   Washington Department of Transportation (WSDOT) 42

APPENDIX A: LIST OF INTERVIEW PARTICIPANTS 47
APPENDIX B: PEER EXCHANGE AGENDA 49
APPENDIX C: PEER EXCHANGE ROUNDTABLE QUESTIONS 51
APPENDIX D: ADDITIONAL RESOURCES 53
APPENDIX E: SELECT STATE DOT APPLICATIONS OF GIS FOR NOISE 54
APPENDIX F: REQUIRED NOISE BARRIER INVENTORY DATA 56

EXECUTIVE SUMMARY

Noise from highway traffic can be pervasive in areas near roadways. How and to what extent noise travels is strongly influenced by geospatial features such as terrain and elevation. Thus geographic information systems (GIS), which enable users to more easily manage, analyze, and present geospatial information, can help transportation agencies evaluate noise impacts from highway traffic and identify noise mitigation options.

There is great potential in applying GIS for highway traffic noise analysis, but its use is still evolving nationwide. To explore how transportation agencies are applying GIS to highway noise analysis, the Federal Highway Administration (FHWA) sponsored a peer exchange on April 23-24, 2012, in Nashville, Tennessee. The peer exchange convened eight State Departments of Transportation (DOTs) that have made progress in using GIS to better understand noise considerations. Prior to the peer exchange, the Volpe Center, in coordination with FHWA, conducted telephone discussions with the peer exchange State DOTs to better understand how agencies are developing GIS tools and applying geospatial data to meet noise program objectives. Case studies summarizing these DOTs' efforts were then developed.

This report, which synthesizes these case studies to describe overall observations, is expected to support GIS and noise practitioners to identify examples of noteworthy practices, consider the pros and cons of GIS applications for noise, and determine how these applications might be best utilized. In general, observations suggest that:

- **State DOTs are using GIS to respond to Federal noise regulations as well as to meet other needs.** Federal noise regulations are codified in Title 23 of the Code of Federal Regulations (CFR), Part 772: Procedures for Abatement of Highway Traffic Noise and Construction Noise. The use of GIS is not required to meet this mandate but many States have chosen to use GIS technologies for this purpose. 23 CFR 772's reporting requirements, which mandate DOTs to provide information on noise walls (e.g., cost, height, length, location) to FHWA on a triennial basis, have been a major driver spurring DOTs in using new GIS technologies for noise.

- **State DOTs are applying geospatial data in different ways to accomplish noise-related objectives.** Participating State DOTs have chosen to develop discrete GIS-based tools focused on noise. Frequently, these tools helped the State respond to 23 CFR 772's reporting requirements. Others are using geospatial data layers available from many different sources to support noise analysis, such as identifying during project scoping when noise barriers might be needed. Typical data sources include State GIS clearinghouses, agency-wide (enterprise) applications, local governments, or online mapping applications like Google Earth.

- **The benefits of using GIS to support noise analysis appear to outweigh the challenges.** State DOTs reported that it was sometimes difficult to find the necessary resources (e.g., staff time and funding) to develop new GIS tools or expand existing ones for noise analysis. Furthermore, very few agencies had formal metrics that assessed the use of current GIS applications. In absence of formal metrics, agencies generally relied on anecdotal evidence to identify the benefits of GIS for noise analysis. These benefits help make the case for supporting GIS for noise analysis and include earlier identification of noise mitigation needs, more effective communication with the public, and time- and cost-savings.

In summary, while the use of GIS for highway traffic noise is still a developing area, State DOTs are currently using a variety of GIS tools and geospatial data from multiple sources to accomplish noise objectives. On the whole, State DOTs believed that these applications, while sometimes challenging to develop and use, led to important benefits such as streamlining responses to Federal mandates and more effective stakeholder communication.

INTRODUCTION

This section provides an overview of the purpose of this research effort and methodology used, as well as details on the Federal noise legislation that provides a basis for agencies' GIS for noise analysis work. This section also provides background information on how applications of GIS for noise analysis fit into the broader context of using GIS to meet transportation needs.

Purpose

Sounds are vibrations in the air that reach the ear. Noise, on the other hand, is unwanted or excessive sound.[1] Unique features associated with a particular space, such as terrain, vegetation, elevation, and land uses, all influence the dynamics of sound, and as an extension, noise. Thus both sound and noise are inherently geospatial issues. Furthermore, highway traffic can be a dominant and pervasive source of noise in both urban and rural environments. GIS, which enables users to more easily manage, analyze, and present geospatial information, can help transportation agencies conduct noise analyses, including assessing how highway traffic will affect noise levels and identifying noise mitigation options.

To explore this topic in depth, Federal Highway Administration (FHWA) and the Volpe National Transportation Systems Center (Volpe Center) conducted a series of telephone discussions and sponsored a peer exchange in Nashville, Tennessee, focusing on select State DOTs' use of GIS in highway noise analysis.

The purpose of the discussions and peer exchange was to allow State DOTs with experience in incorporating GIS into highway noise applications and activities the opportunity to:
- Identify the state of the practice in applying GIS to noise analysis;
- Share related experiences, including technical approaches and innovative examples;
- Discuss and document benefits, challenges, and lessons learned;
- Identify strategies to expand the use of GIS in noise analysis; and
- Support such an expansion through networking and identifying a GIS for noise community of practice.

The use of GIS in noise analysis is relatively new. This effort convened noise and GIS specialists to discuss the burgeoning intersection of GIS and noise expertise, a dialogue that may not have previously occurred on a regular basis among the selected States. Furthermore, in light of a July 2011 update of Federal noise regulations, the interviews and peer exchange also provided an opportunity for practitioners to discuss new approaches for using GIS to address Federal mandates.

By hosting peer exchanges and supporting other research and activities focused on a variety of topics, FHWA seeks to facilitate opportunities for States to gain knowledge from others' successes and challenges in the application of GIS.[2] Furthermore, FHWA shares resources that can support agencies' use of GIS (see Appendix D for a list of resources).

Methodology

State DOTs were selected for participation based on their activity as part of the Transportation Research Board's Committee on Transportation-Related Noise and Vibration (ADC40). While the selected State DOTs varied widely in their approaches to using GIS to support noise analysis, all had demonstrated progress in this area.

---

[1] FHWA. Highway Traffic Noise in the United States: Problem and Response. 2006. www.fhwa.dot.gov/environment/noise/regulations_and_guidance/usprbrsp.pdf

[2] Additional information on FHWA's efforts to promote transportation agencies' applications of GIS, geospatial tools, and data can be found on the FHWA GIS in Transportation website: http://gis.fhwa.dot.gov/.

Participants included noise and/or GIS staff from the California Department of Transportation (Caltrans), Florida DOT (FDOT), Maryland State Highway Administration (MDSHA), North Carolina DOT (NCDOT), Ohio DOT (ODOT), Tennessee DOT (TDOT), Virginia DOT (VDOT), and Washington DOT (WSDOT). Appendix A includes a complete list of participants. An interview guide provided a framework for the telephone discussions conducted prior to the peer exchange (see Appendix B for the guide). Each discussion lasted approximately 60 to 90 minutes. Case studies were drafted based on participant responses during these discussions. A set of draft case studies was distributed to participants as background material ahead of the peer exchange.

The peer exchange, which TDOT hosted at its offices in downtown Nashville, took place on April 23-24, 2012. During the peer exchange, FHWA presented its perspectives on the benefits of applying GIS to noise analysis. FHWA believes that GIS, use of geospatial data, and other geospatial tools (e.g., remote sensing) can help transportation officials make better decisions.

Following FHWA's presentation during the peer exchange, State DOTs demonstrated their current uses of GIS and geospatial data to support their respective noise analyses. Four roundtable discussions also provided opportunities for dialogue. These roundtables focused on topics of interest identified from the telephone discussions, including: (1) using GIS for noise reporting (sound wall inventories); (2) collecting noise-relevant data; (3) applying GIS/geospatial data to assess noise; and (4) assessing the potential of GIS in noise assessments. Appendix C provides a list of questions discussed during the roundtables.

Appendix E provides a cumulative look at all applications of GIS discussed during the telephone discussions and peer exchange as well as how these applications are used.

Background

The Federal-Aid Highway Act of 1970 includes mandates for how FHWA must regulate and mitigate highway traffic noise. FHWA codified its regulations in Title 23 CFR, Part 772: Procedures for Abatement of Highway Traffic Noise and Construction Noise.[3] The regulation, which applies to any highway project that receives Federal funding, was updated in July 2011 to provide additional guidance that reflected feedback received from State DOTs and professional practitioners.

The regulation also updated definitions of key noise terms. Terms used frequently in this report include the following:

- **Receptor:** a discrete or representative location of a noise sensitive area(s) (e.g., hospital, residence, park, school).

- **Noise barrier:** a physical obstruction that lowers the noise level and is constructed between the source and receiver of highway noise.[4]

- **Type I project:** a noise abatement project that the National Environmental Policy Act (NEPA) requires State DOTs to undertake when constructing new highways or making substantial highway alterations or additions.

Highlights of July 2011 Updates to 23 CFR 772:
- Required analysis for an expanded list of Type I projects;
- Required use of Traffic Noise Model (TNM) version 2.5;
- Added a requirement to report noise barrier inventories to FHWA on a triennial basis (the "reporting requirement"); and
- Requirement to notify local officials about future noise levels and whether a community is eligible for a Type II noise program ("noise compatible planning").

---

[3] For the complete text of 23 CFR 772, http://www.ecfr.gov/cgi-bin/text-idx?tpl=/ecfrbrowse/Title23/23cfr772_main_02.tpl.
[4] "Noise wall" is used here interchangeably with "sound wall" or "noise barrier." Participants in this research effort regularly used all of these terms.

- **Type II project:** a voluntary noise abatement project that a State DOT undertakes to add noise abatement measures to existing highways (10 State DOTs currently have a Type II program).

The update of 23 CFR 772 formalized reporting requirements. State DOTs must provide to FHWA (on a triennial basis) inventories that include information on any constructed noise abatement measure, including cost, height, length, location, year of construction, and materials used in construction. See Appendix F for the complete list of data required in these noise wall inventories.

Furthermore, DOTs must use FHWA's TNM in their traffic noise prediction analysis for all projects subject to 23 CFR 772. The FHWA TNM is a state-of-the-art, three-dimensional model that calculates traffic noise levels and noise level reductions based on user input of roadways, barriers, terrain features such as hills, valleys, woods and lakes; structures, and traffic data. It uses advances in personal computer hardware and software to improve upon the accuracy and ease of modeling highway noise, including the design of effective, cost efficient highway noise barriers. Modeling a location requires a significant amount of data including topography, the physical characteristics of the existing noise barrier, traffic volumes, and mix and speed of the adjacent highway.[5]

While neither 23 CFR 772 nor TNM require the use of GIS, many States have turned to GIS to facilitate compliance with 23 CFR 772 and use of TNM, as well as to support noise analyses.

The capabilities of GIS in the transportation arena are well documented.[6] However, there is still unexplored potential for using GIS to support highway traffic noise analysis. Existing research primarily focuses on technical requirements associated with GIS-based noise modeling.[7] There is currently little research on how GIS can support other noise activities, such as in compiling noise wall inventories. Little documentation exists on the lessons learned, success factors, and challenges experienced by transportation agencies using GIS to address highway traffic noise, as well as the steps agencies took in developing discrete GIS tools focused on noise.

Most of the agencies participating in this research effort are still investigating how to most effectively develop GIS tools for noise or utilize geospatial data for noise analysis. Resources for noise programs can be limited, complicating agencies' efforts to initiate or expand their uses of GIS in the noise discipline. Typically, noise programs at State DOTs are comprised of fewer than three staff persons and many programs have worked extensively with consultants to collect noise-related geospatial data. GIS and noise specialists within the same agency may not have regular opportunities to coordinate and discuss GIS solutions to support noise analysis.

Despite these challenges, some State DOTs have found innovative ways to use GIS and geospatial data, finding that GIS can support many aspects of noise analysis, including meeting the requirements of 23 CFR 772 (particularly its reporting requirements). As agencies experience successes in these areas, these accomplishments will likely provide good foundations for subsequent advancements.

---

[5] Additional information about TNM is available at www.fhwa.dot.gov/environment/noise/index.htm.
[6] For many examples of how GIS has been used to support an array of transportation goals and objectives, see the FHWA GIS in Transportation website.
[7] For some examples of this research, see "Simple Tools for Traffic and Transit Noise Studies." 2002), available at http://proceedings.esri.com/library/userconf/proc02/pap0505/p0505.htm. See also "GIS-Based Model for Noise Analysis" (2009), available at http://ascelibrary.org/iso/resource/1/jitse4/v15/i2/p88_s1?isAuthorized=no.

OBSERVATIONS

This section describes overall observations in how select State DOTs are using GIS to analyze the potential effects and impacts of highway noise, as well as the associated benefits, challenges, and lessons learned resulting from these experiences.

State of the Practice

The use of GIS to support noise analysis is evolving nationwide. The recent update to 23 CFR 772 has spurred some State DOTs to initiate or re-consider the use of better or emerging GIS technologies to access geospatial data in new ways than in the past.

- **Use of GIS for noise analysis is still a relatively new activity among State DOTs.** Many State DOTs are in the early stages of developing formal GIS tools for noise analysis. Most have not had extensive opportunities in the past to foster coordination between GIS and noise specialists. All of the State DOTs participating in this effort noted that they plan to expand or modify existing GIS noise tools or else develop new GIS for noise analysis solutions in the future.

- **States are using both geospatial data and discrete GIS tools to accomplish noise program objectives.** All participating State DOTs reported looking to a variety of sources to obtain geospatial data to support noise analysis. Important noise-related information includes topography, elevation, traffic volumes, and land uses. Frequently used sources for these and other data include agency or State enterprise GIS systems, counties and cities, national sources such as the U.S. Geological Survey (USGS), and online tools such as Google Maps and Bing Maps. In some cases, State DOTs have compiled data to develop discrete GIS tools focused on noise (typically, the initial purpose of these discrete tools was to respond to the inventory requirements of 23 CFR 772). In a few cases, however, State DOTs did not have discrete tools for noise analysis although they planned to move in this direction in the future. These agencies are continuing to rely on accessing noise-related geospatial data from multiple sources to support noise analyses.

- **GIS can help meet the requirements of 23 CFR 772 as well as other noise program objectives.** With the exception of NCDOT, all other participating State DOTs are using or are starting to use GIS applications to respond to 23 CFR 772's reporting requirements. Specifically, some State DOTs are using GIS to support identification of potential noise impacts during project screening and scoping phases of project delivery. For example, with its Environmental Screening Tool (EST), a web-based GIS application, FDOT and other users can access noise-related information (in conjunction with other environmental data) to assess the potential impacts of planned transportation projects.

    No State DOT had developed a discrete GIS tool to meet 23 CFR 772's requirement on coordinating with local officials, although a few DOTs mentioned that they are currently working or plan to work closely with local governments to collect noise-related geospatial data. Through this coordination, some State DOTs are engaging in some aspects of noise-compatible planning. However, participants noted that these discussions are typically advisory in nature since the State DOT does not have the authority to develop and implement local land use plans.

    Beyond meeting specific 23 CFR 772 requirements, some State DOTs are using GIS to support noise decision-making (e.g., regarding maintenance schedules). State DOTs are also using GIS to support communication both within and outside the agency, particularly in regards to addressing the public's noise barrier inquiries.

    Additionally, though with somewhat lesser frequency, State DOTs are using GIS to support traffic model development and design functions, often as part of TNM. NCDOT has also begun using geospatial data to design noise barriers. Finally, several State DOTs have found value in using web-based GIS tools to facilitate communication with the public, particularly in regards to noise barrier issues and complaints, such as complaint tracking and identifying complaint hot-spots.

- **States have used GIS for noise analysis to meet different needs, but their uses have changed over time.** Most participating State DOTs found that uses of GIS tools for noise analysis evolved as different needs emerged. For example, TDOT began using GIS to identify potential locations of Type II barriers in response to the TDOT Commissioner's request for a Type II study. TDOT is now planning to modify its noise wall GIS inventory to generate the required inventory reports.

- **Measures to assess the effectiveness of GIS tools for noise analysis are not yet common.** Some State DOTs are using anecdotal evidence or limited quantitative data to evaluate their use of GIS for noise analysis. For example, WSDOT collects statistics on usage of all datasets in the GIS Workbench, an agency-wide GIS application containing environmental and natural resource data (WSDOT tracks the number of users and the number of times the datasets were accessed). However, no State DOT reported having formal metrics that evaluate whether tools are meeting users' needs or to measure cost or time savings. This could be due to the fact that GIS use in noise analysis is still a relatively new practice, and State DOTs have not yet obtained sufficient experience or data for comprehensive evaluation. As GIS tools for noise analysis mature, it is expected that agencies will be able to develop more robust performance measures.

- **Use of GIS to communicate with the public on noise could be expanded.** Some State DOTs are using GIS for noise applications to communicate with the public, but are doing so in somewhat limited ways. In general, this communication has focused on geocoding noise complaints, for example pinpointing locations where individuals have identified problems with noise walls or have suggested the need for a new noise wall; responding to public inquiries about existing noise walls; and depicting noise contours at public meetings.

Benefits

State DOTs reported a number of benefits related to uses of GIS for noise analysis, particularly in terms of streamlining communication with diverse stakeholders, making noise analyses more efficient and cost-effective, supporting better coordination, and fostering more effective decision-making.

- **GIS streamlines communication on noise to both internal and external stakeholders.** Nearly all State DOTs reported that use of GIS for noise improved communications. This was particularly true for clearinghouse-type tools that compiled an array of noise information into "one-stop shops" that helped cement institutional knowledge about an agency's noise activities. As repositories, these tools helped staff members respond more quickly and thoroughly to inquiries from agency leadership and the public. Publicly accessible tools, such as FDOT's EST, also provided more direct ways for the public to engage in noise discussions with a State DOT and understand the actual or potential noise impacts of transportation projects. Use of GIS for public communication has potential to grow; however, well-defined practices, such as what information is appropriate to share, have not yet been developed.

- **Use of GIS for noise analysis can save time and costs.** While State DOTs had not developed formal measures to assess the extent to which GIS tools for noise analysis provided time- and cost-savings, many reported anecdotal evidence suggesting these benefits. For example, NCDOT has very accurate elevation data. Because of this, the agency has been able to design noise walls and berms using GIS without the need for an accompanying field survey, although it still conducts surveys prior to construction. This has avoided survey redundancies, made it easier for NCDOT staff members to meet project deadlines, and saved both time and costs. FDOT also noted that considering noise as part of project screening phases has led to better project outcomes.

- **GIS tools can support more strategic, coordinated agency decision-making regarding highway noise.** Some State DOTs noted that in establishing more organized information repositories, GIS tools helped staff make better decisions about noise walls. For example, Caltrans reported that its online GIS noise wall inventory allowed staff members to better assess where it was most cost-effective to build a noise wall. Through its GIS Workbench application and other companion tools, WSDOT has also fostered more coordinated decision-making as staff members in different divisions can identify how noise walls might affect their areas of project planning and development.

- **GIS enables more comprehensive, accurate traffic noise modeling.** Four State DOTs (NCDOT, ODOT, TDOT, and VDOT) reported that they are using geospatial data in conjunction with TNM. These DOTs believed that doing so had led to a more comprehensive understanding of noise impacts. For example, NCDOT uses GIS data with TNM to capture additional detail on receptor features and land uses outside the highway right of way, enabling more finely tuned traffic noise level prediction. Generally, the use of GIS as part of noise modeling, with or without TNM, can help staff members assess impacts on a larger number of receivers, present results in map formats that might be easier for the public and others to understand, and more quickly and easily remodel if necessary.[8]

Challenges

State DOTs have demonstrated creativity in developing a variety of innovative GIS applications and tools for noise analysis. However, given the nascent nature of the state of the practice, State DOTs have experienced some challenges in developing these activities, including:

- **Gathering noise-related GIS data can be difficult.** Some State DOTs reported that they must "cobble together" data from multiple sources for noise analyses, which in some cases added time and cost. This was true even for State DOTs that developed discrete GIS tools for noise or that had access to enterprise systems. For example, ODOT has developed a GIS noise barrier tool but continues to use information from Google Earth to get the information it needs to conduct noise modeling.

  Having data on areas beyond the highway right of way is important in noise analysis to capture information on noise receivers. However, obtaining these data can be difficult since engineers conducting roadway field surveys rarely collect them. Local governments can provide these data, but this requires additional coordination and sometimes costs.[9] Additionally, the data files occasionally have inconsistencies with State DOT data (e.g., in terms of scales or resolution levels), making it difficult to work with these data without first standardizing them.

  Finally, some State DOTs reported challenges in obtaining information on "as-built," or constructed, noise walls in order to update noise barrier inventories, regardless of whether these data were in GIS format. This is because older as-built plans might not be easily accessible, or because contracts developed between State DOTs and contractors who actually construct the noise barriers often do not contain stipulations to submit final as-built plans. Even when these stipulations are included, they may not be consistently enforced and delivered information may not be consistently managed. Further, there is often limited or no funding available to support updating GIS databases with as-built information. Collecting as-built data is important because a constructed barrier can differ from the initial design in terms of elevation profiles, alignment and offset, and other features such as material, color, and surface treatments. An inventory that contains information only from designed barriers might not be accurate. Without an easy way to

---

[8] "Simple Tools for Traffic and Transit Noise Studies" (2002) is available at http://proceedings.esri.com/library/userconf/proc02/pap0505/p0505.htm.
[9] North Carolina's "NC One Map," a statewide repository for local-level natural resource data, served as an exception, as NCDOT noted that access to local-level data through this tool is continually being expanded.

obtain as-built data, staff members must collect this information in the field, which adds time and cost.

- **Using GIS in conjunction with TNM requires workarounds.** The TNM currently accommodates import of geospatial data through a tool that can import two or three dimensional drawing exchange format files. However, TNM does not currently offer a direct interface with GIS. As a result, State DOTs wanting to use GIS with TNM must take additional steps to do so, such as converting data into appropriate file formats. Some State DOTs emphasized that even when GIS data are used with TNM, it is important to remember that TNM's primary function is to calculate noise levels. Achieving accurate noise modeling is related to modelers' skills and the quality and detail of inputs as opposed to whether or not these inputs are in GIS format. TNM version 3.0 is expected to contain features that make integration with geospatial data easier.

- **DOT noise programs often have limited resources.** Most State DOTs participating in this effort had relatively few staff members (either GIS or noise specialists). Some reported difficulties finding available staff and funding to dedicate time to GIS for noise analysis, particularly data collection (for details on data useful for noise applications, see also Tables 1 and 2). State DOTs have addressed these difficulties by working with consultants to augment staffs' work, finding low-cost solutions to collect data, and leveraging existing work rather than beginning efforts "from scratch." For example, TDOT plans to make modifications to an existing GIS database to allow development of the Federally-required noise wall inventory.

- **Articulating the value of using GIS for noise analysis can be difficult.** Some State DOTs believed that agency leadership did not fully understand the benefits of GIS or how its application in noise analyses could meet other, broader agency needs. As a result, staff members have often felt limited in their ability to be proactive, particularly when reaching out to others in the agency to obtain data or coordinate in other ways.

- **Public information on noise issues needs to be carefully developed.** 23 CFR 772 requires State DOTs to share information on estimated noise levels from projects with local officials. State DOTs expressed concern about using the noise contours generated by TNM to convey this information, since the contours could be misinterpreted as actual impact assessments rather than early screening tools; TNM 3.0 will depict noise levels as gradients rather than contours to help mitigate this problem. State DOTs generally agreed that as the availability of publically accessible GIS tools increases over time, the information provided needs to be balanced in order to manage expectations and prevent misunderstandings.

Potential Advancements for GIS in Highway Traffic Noise Analysis

As State DOTs discussed the future of their noise programs, they commonly noted several promising and desired advancements that geospatial data and tools could achieve within the practice of noise analysis, including:

- **GIS could support early screening for noise barriers.** GIS and geospatial data convey detailed highway information. With this information more readily accessible, noise practitioners may be able to better predict when project alternatives will likely need a noise barrier. Conducting these analyses early in project planning and scoping phases can save transportation agencies time and money by allowing project managers to proactively respond to abatement needs through design solutions or budgets.

- **GIS could help in the design of noise barriers.** NCDOT has successfully used GIS to design noise wall top of wall (acoustic) profiles to achieve desired decibel reductions. NCDOT has the advantage of readily accessible light detection and ranging (LiDAR) data and statewide natural resource data, but notes that with similar resources, other State DOTs could save time by designing some elements of noise barriers using GIS before waiting for field survey data.

- **The TNM could be more compatible with GIS.** A recurring concern that State DOTs expressed was that using geospatial data in conjunction with TNM for noise modeling was difficult. The next version of TNM (3.0) will offer features to enable easier integration with GIS. In the meantime, there might be opportunities to identify how State DOTs can use GIS more effectively as part of the current version of TNM (2.5).

- **Noise barrier inventories could be used to better anticipate maintenance needs.** State DOTs recognized that many constructed noise barriers are now reaching the end of their design lives, so there is little precedent for how to anticipate or address maintenance issues. However, State DOTs learned that GIS-based noise barrier inventories offer opportunities to better understand not only the locations, height, and cost of the noise walls, but their conditions as well. Going forward, GIS offers the ability to better track maintenance needs related to noise barriers and build a historical record that can be used to predict future needs based on factors like age, location, and environmental conditions.

Lessons Learned

State DOTs that have used GIS for noise analysis indicated that their efforts typically began in a piecemeal fashion, applying GIS and geospatial data to address various issues as they arose and gradually expanding GIS use. In hindsight, State DOTs have identified both successful approaches and missed opportunities to help inform the further integration of GIS. Lessons learned include:

- **Be strategic.** It is important to be strategic and thoughtful about how GIS will support noise analysis before investing significant time and costs in developing geospatial applications and datasets. Planning ahead will reduce potential compatibility issues with other platforms or data and information systems, as well as limit redundancies and missed opportunities in terms of data collection and application design. The need and purpose of GIS tools for noise analysis should be clarified, articulated, and prioritized at the outset. Potential or future needs should also be considered.

- **Make a business case for GIS for noise analysis.** Some State DOTs believed that noise programs had insufficient resources and that too few opportunities existed to coordinate with staff members from other divisions or program areas within their respective agencies. Without broad agency support, the full value of GIS for noise analysis will likely be limited. To develop the business case for GIS for noise analysis, State DOTs reported successes in tying GIS to higher-level agency goals such as delivering cost-effective transportation projects and increasing customer satisfaction. Agencies should articulate how GIS leads to better transportation outcomes, instead of focusing only its ability to store or warehouse information. For example, NCDOT actively articulates its use of GIS as a way to potentially save the agency millions of dollars in unnecessary noise walls or lawsuits through contributing to more accurate noise modeling. MDSHA developed an enterprise GIS system called MD iMAP, which provides a common data formatting structure throughout the agency.[10] MDSHA noise and GIS staffs saw the value in adding noise data to a system that leadership already supported. Currently, staff members are formatting noise data for use on MD iMap.

- **Tailor geospatial datasets and GIS tools for noise analysis to State DOT needs.** State DOTs recognized that there is currently no "best practice" in terms of the type or quality of GIS data, outputs of noise applications, or design of tools and applications to meet noise objectives. The variety of GIS applications highlighted in this effort served as evidence that GIS can be highly customized to match unique needs, goals, and resources.

- **Identify what data are needed and where information can be obtained.** When developing GIS tools for noise analysis, a starting point for obtaining noise-related information could be reaching out to staff members in other programs or divisions within an agency. Information useful for noise

---

[10] Additional information on MD iMap is available at www.imap.maryland.gov/portal/.

analysis can also be collected from existing GIS tools or applications, particularly those focused on land uses or environmental or natural resource management (see Table 1). LiDAR and topographic data could also enhance the accuracy and versatility of geospatial noise applications. Transportation agencies often produce detailed computer-aided design (CAD) files of terrain, but many agency staff members might not be aware of the availability of these files. Table 2 provides some examples of useful data to support certain noise analysis activities. Data could be used for multiple activities, including others not listed.

State DOTs should consider what older data are essential, since it can be a very time-intensive activity to collect historical information from various sources that might be in inconsistent formats.

Table 1. Useful Sources of Data for GIS Noise Applications.

State LiDAR data
CAD topographic data
Pavement assessments
Maintenance inventories
Statewide parcel or natural resource data

Table 2. Important Data for GIS Applications to Noise Analysis.

Activity	Important Data
TNM or Noise Modeling	Ground elevation Data on the "top of the wall" elevations Location of structures Vegetation Land cover Drainage patterns Utilities
Project Scoping / Screening	Land uses categorized by noise impact threshold (as defined in 23 CFR 772)
Developing a Noise Wall Inventory	Data on the "top of the wall" elevations Manufacturer of noise wall materials Other information as required by 23 CFR 772
Public Communication	Type I and II barrier locations

- **Collaborate with relevant partners from the beginning.** State DOTs noted that collaboration with partners is key to ensuring that GIS tools for noise analysis are successfully implemented. For example, just after ODOT designed a GIS-based viewer for its noise barriers, their information technology (IT) department implemented a new system design that prevented the centralized use of the viewer across districts, leading to ODOT abandoning the tool. From the outset, agencies should work closely with their IT departments, as well as others throughout the agency, to ensure that any GIS tool developed will be accessible and compatible with existing technologies and platforms. Through this collaboration, staff members might learn that others have a need for noise-related information, which can serve to expand the customer base for GIS applications and data.

- **Allow for flexibility.** State DOTs recommended that geospatial data and tools should be valued for their dynamism and new uses should be encouraged to emerge and evolve over time, including repurposing old tools. While standards and procedures are important, they should not be so constrained as to lead to stagnation. For example, ODOT was able to transform its noise barrier inventory from a storehouse of maintenance information into a more comprehensive tool to manage noise complaints. New customers can also be identified for existing tools. Over time, GIS experience will likely lead to new ideas for generating previously unavailable information. NCDOT, for example, proposed to integrate noise barrier locations with environmental and maintenance data as a way to better predict their lifespans.

- **Streamline data collection and formatting.** State DOTs identified several potential strategies to ensure seamless integration of data from disparate sources into GIS noise tools and applications moving forward.
    - Develop a consistent, systematic approach for obtaining and georeferencing information.
    - Identify procedures for collecting data on as-built walls, including an appropriate submission schedule.
    - Determine a format, system, and/or regular schedule for how noise data are added to a GIS tool so that all users know they are accessing up-to-date information.
    - Coordinate with others throughout the agency to capture additional data that might assist with noise analysis. Inform others if noise data or tools are accessible for use.
    - Identify what resolution and scale are required for data.
    - If using GIS data with TNM, consider saving all model inputs so that they can be easily replicated if needed, such as if a road is widened.

- **Be resourceful.** Most participating State DOTs acknowledged that they were working with limited staff and financial resources as they embarked on using GIS in their noise programs. State DOTs indicated having utilized the following approaches when challenged by limited financial resources or staff capacity:
    - "Key off" existing efforts to leverage resources. For example, VDOT noise staff collected noise wall data (e.g., wall start and end points, construction materials, and conditions) as part of a pavement assessment effort that a consultant was already conducting for the agency.
    - Partner with academic institutions or utilize statewide geospatial data repositories, if available. For example, FDOT established a partnership with the University of Florida's Geoplan Center to maintain data used as part of the EST.
    - Identify free online tools to support data collection, if possible given agency policies. Many of these tools can also be used to link to "street view" visuals. Caltrans needed only one staff member to build a customized ESRI FlexView viewer for its noise program. State DOTs cautioned that the source of the geospatial data for these tools should be verified and validated.
    - Work with consultants when necessary to augment staff members' work.

Areas for FHWA Support

In addition to the above lessons learned, State DOTs mentioned several ways in which FHWA could support efforts to incorporate GIS and geospatial data into noise analysis activities.

- Develop and distribute a template for the level of detail for data that should be collected for noise barrier inventories. Reported information could be scaled or customized according to a State DOT's needs. MDSHA suggested consulting the U.S. Department of Defense's Spatial Data Standards for Facilities, Infrastructure, and the Environment[11] for potential models for this template.
- Standardize how components of noise barriers are defined and should be organized in noise barrier inventories submitted to FHWA. State DOTs reported difficulties in determining how noise barriers units should be recorded if there are breaks or significant elevation changes in the barrier (e.g., if a barrier encounters a bridge).
- FHWA could conduct additional outreach to State transportation agencies to support them in making the business case for using geospatial data and tools in noise analyses.
- FHWA could research potential coordinate systems to replace the current state plane systems, which State DOTs noted were generally only useful to surveyors and within small areas.

---

[11] Available at http://www.sdsfie.org/.

CASE STUDIES

This section presents in-depth case studies on the current activities of the State DOTs that participated in interviews and the peer exchange. Each case study includes information on how the agency began using GIS or geospatial data to support noise analysis activities, how it developed noise analysis GIS tools, and the challenges, lessons learned, and benefits encountered while utilizing their respective tools.

California Department of Transportation (Caltrans)

Background

California has an estimated 750 miles of noise walls. Due to the large number of noise walls distributed across such a large State, Caltrans sometimes struggled to coordinate and maintain its sound wall inventory among its 12 districts and 22,000 employees.

> Caltrans used GIS to develop a web-based "Statewide Sound Wall Inventory" and mapping application.

To help address this issue, Caltrans identified GIS in 2009 as a way to incorporate a large range of data. Using GIS, Caltrans compiled sound wall information into a GIS-based inventory. Information from the inventory was previously compiled in Excel spreadsheets. The overall intent of developing the inventory was to respond to the reporting requirements of 23 CFR 772, but over time the tool has helped support broader decision-making related to the sound walls.

The inventory was converted into an interactive, web-based tool that assists users in visualizing noise walls and any issues associated with them. The tool is accessible to all Caltrans' districts and the public at http://www.dot.ca.gov/hq/env/noise/index.htm (see Figure 1).

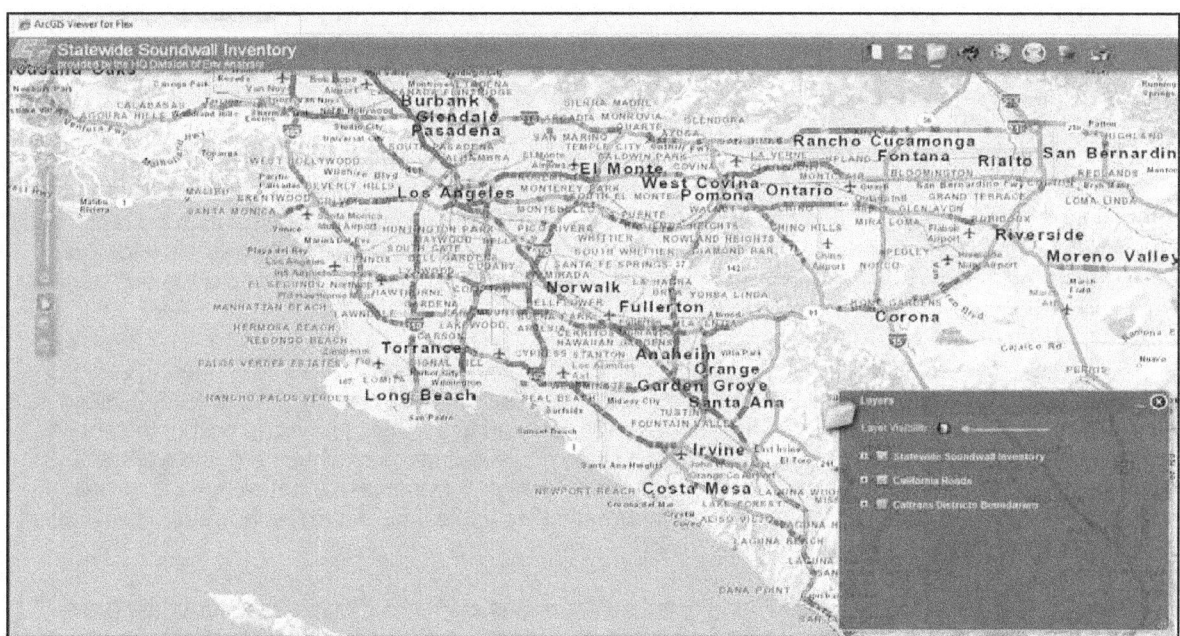

Figure 1. Screenshot of Caltrans' Online Statewide Sound Wall Inventory.

Developing the Sound Wall Inventory

The overall purpose of the inventory was to provide a "one-stop shop" for obtaining information on the location of sound walls and their attributes, including their length, location, and construction costs. The inventory also provides users with an easy way to visualize noise walls and conduct queries. Because Caltrans' districts each have their own GIS applications, some of which include noise data, there is no centralized repository that stores all of Caltrans' GIS noise data. In developing the inventory, Caltrans sought to provide a more consistent way to store and communicate information on noise walls.

The inventory, which one staff member built, uses ESRI's FlexViewer template, a tool available for free online and designed to be used in conjunction with an ArcGIS server.[12] FlexViewer is customizable, user-friendly software that enables users to create online mapping applications.

To develop the tool, Caltrans staff georeferenced information from Excel-based spreadsheets at a cost of approximately $12,000 including the cost of GIS software, maintenance fees, and labor. Additional information was obtained through requests to Caltrans' districts, Google Maps, or from site visits by district staff when possible. Caltrans's existing GIS-based maps (streets, aerials, and topography) were also imported into the tool. Type II sound walls were not included since they are restricted to local funding.

In addition to sound wall data, the inventory also contains air quality and school data. The latter are included due to a State regulation that transportation project noise levels in proximity to schools must be minimized.

Noise wall attributes within the inventory are linked to QuickTime videos of five-mile sections of the barriers. These videos were created as part of the Caltrans' previously conducted video inventory of all State highways. Adding video to the inventory took a few weeks.

Currently, the inventory is referenced when staff members respond to noise complaints or when existing walls need to be identified, for instance, when adding lanes to a highway. While Caltrans does not yet utilize GIS in conjunction with TNM, Caltrans notes the inventory will be useful in determining the elevation of sound walls as part of noise modeling for paving or other highway expansion projects. The inventory is not currently being used during project screening or scoping phases.

Sound Wall Inventory Data

Caltrans recognizes that there are some information gaps in the existing inventory and has estimated data points for some attributes, such as wall length and location. Many of the data gaps are related to attributes for as-built walls. This is because older as-built information was discarded and there is currently no system in place for Caltrans to automatically receive as-built information from its districts. However, Caltrans is working on developing a system to obtain elevations for constructed sound walls and heights from districts once projects are completed.

Wall cost data are also difficult to obtain due to California's procurement process. Engineering estimates provide a range for wall costs, but bids for noise walls are made based on sound wall material unit costs, often for multiple walls at once. Yet, what is needed to respond to 23 CFR 772 is the cost of the sound wall per square foot. To date, Caltrans has been able to make estimates of what the noise walls actually cost to construct by comparing bids to actual unit costs, though it recognizes a future need for more accurate noise wall cost data. Collecting such information would be time-intensive, however, as it would require searching each district's database.

In the future, Caltrans would like to include noise contours and cumulative noise exposure data from airports in the inventory, should these be developed (the city of San Francisco has such information, including stationary sources with contours). Caltrans would ideally like to have contour data from across the State with 500' intervals; however, such a large undertaking is unlikely given current funding and staffing constraints. As an alternative, some cities may be willing to provide Caltrans with their noise data.

---

[12] For more information on FlexViewer, see http://help.arcgis.com/en/webapps/flexviewer/help/index.html.

Another potential use of the inventory could be determining how different pavement types used for highway rehabilitations will affect noise impacts, but these data are currently not included in the inventory. Caltrans views pavement types as a desirable, though potentially difficult to obtain, addition to its GIS database. Caltrans also intends to explore whether LiDAR data might be available within its organization to help further verify existing sound wall data.

Benefits

Overall, the inventory's benefit to Caltrans is in allowing staff to conduct more accurate and efficient spatial analyses and providing a more comprehensive understanding of what walls exist and where they are located. This information supports decision-making on noise abatement measures, such as identifying where building a wall will be most cost-effective. Caltrans has also found the tool useful to obtain elevation information.

The inventory also supports more effective information sharing. A selected view or extent can be bookmarked and shared with another user. Users can also perform detailed queries; mousing over results will reveal an attribute table that contains more information on the noise walls. Each attribute table contains a link to a Google map from which users can select a "street view" if desired. The Google map is also embedded with a marker that will launch the corresponding video of the road when clicked. Data within the tool can be saved locally on a computer, or exported for use in reports.

Lessons Learned

Caltrans identified the following lessons learned in developing its GIS noise wall inventory:

- **Determine in advance the needs the application will address and its potential uses.** If users simply need to identify noise wall locations or their height and length, then data from Google Maps or other mapping applications might suffice. However, for more complex queries or developing detailed maps, agencies should consider building a more comprehensive GIS tool that provides access to a broader array of information. Caltrans also found that ESRI's online tutorials and resources[13] were especially useful in learning how to develop and build a web-based, GIS-based application.

- **Take a broader view of the effectiveness of GIS applications.** Since coding is a focused process, Caltrans reported that it could be initially difficult and relatively costly to dedicate staff time to building a GIS application. On the other hand, Caltrans found that once developed, these applications, as well as geospatial data available from online mapping applications such as Google Earth, could provide significant time-savings over the longer term. For example, using Google Earth, staff members can look online for information as opposed to needing to drive to a location to survey land use features close to noise walls.

---

[13] More information is available at http://help.arcgis.com or http://help.arcgis.com/en/arcgisexplorer/help/.

Florida Department of Transportation (FDOT)

Background

In 2001, FDOT and 23 other State, Federal, and local agencies, as well as several Tribal Nations developed the Efficient Transportation Decision Making (ETDM) process. These agencies and Tribal governments form Environmental Technical Advisory Teams (ETAT) for each of FDOT's seven geographic districts. ETDM provides a framework for FDOT, the ETAT, and the public to identify potential environmental impacts of transportation projects beginning at the earliest planning stages. The ETDM process promotes early coordination and transparent communication that leads to better issue definition, leading to a reduction in the amount of unforeseen, late-stage events that can affect project schedules and budgets.

> **FDOT uses GIS in the noise discipline to:**
> - **Coordinate noise assessments and analyses with other agencies through an environmental screening tool.**
> - **Inventory noise walls and vibration-sensitive sites.**

FDOT's Central Environmental Management Office provides oversight of FDOT's noise program in regards to rules and procedures for all noise analyses in the State. Administrators in each of FDOT's seven districts and turnpike authority oversee the actual implementation of all projects located within their areas. Each of FDOT's Districts has a noise specialist who works with its own consultant team to complete and review noise assessments.

The ETDM process is implemented using the EST, a web-based, GIS application[14] that contains all comments and information related to screened projects and over 550 environmental resource GIS data layers, including over 25 that are noise-related. ETATs review and comment on proposed transportation projects within the EST according to their agencies' respective statutory and regulatory authority. The public can also access the EST and use the tool to stay updated on projects of interest.

FDOT and environmental agencies use the EST to address noise issues early on in the project planning and programming phases. Noise is one of 21 issues considered during these phases, with noise impacts reviewed in regards to their relationship with population centers, land use, proximity to existing noise walls, and proximity to noise sensitive facilities like hospitals. Standard noise modeling analyses are conducted on transportation projects after the planning and programming phases, during the project development and environment (PD&E) phase. Currently, FDOT does not incorporate geospatial data into the TNM process as part of PD&E.

FDOT is in the early stages of updating its GIS noise wall inventory to respond to 23 CFR 772's reporting requirements and other internal and external requests for noise wall information. All GIS layers are stored within the Florida Geographic Data Library (FGDL), a comprehensive database at the University of Florida's GeoPlan Center.

Assessing Noise in the ETDM Process

Given the impact noise walls have on project costs and schedules, FDOT saw value in considering noise early in the project delivery process. Prior to 2012, however, noise considerations and comments were provided under EST's aesthetic issue. Under this arrangement, however, noise was not always fully addressed, so FDOT decided to treat noise as a stand-alone issue within the EST to better highlight its importance. Noise impact considerations within the EST primarily consider the community's perception of possible noise impacts because project alignment details at this stage are still too fluid to conduct an analysis in terms of physical impacts. Predictive modeling does not take place until later stages.

---

[14] More information on ETDM and the EST is available at https://etdmpub.fla-etat.org/est/.

The ETDM process consists of three phases: planning, programming, and project development and environment (PD&E) (see Figure 2). During the Planning and Programming Phases, projects are reviewed in the EST. The reviews, also known as screening events, last 45 to 60 days and are initiated through EST-generated email notifications. The Planning and Programming Screens apply only to qualifying capacity improvement projects as outlined in the ETDM Manual and include such projects as the widening of roadways, new roadways, new rail systems, and bridge projects. The issues that agencies are expected to comment on during each screening are specified in the Agency Operating Agreement that exists between the agency, FDOT, and FHWA. FDOT and metropolitan planning organizations (MPOs) are responsible for sociocultural effects evaluations. FDOT, MPOs, and FHWA are the only agencies expected to comment on noise issues, though other agencies may elect to comment as well.

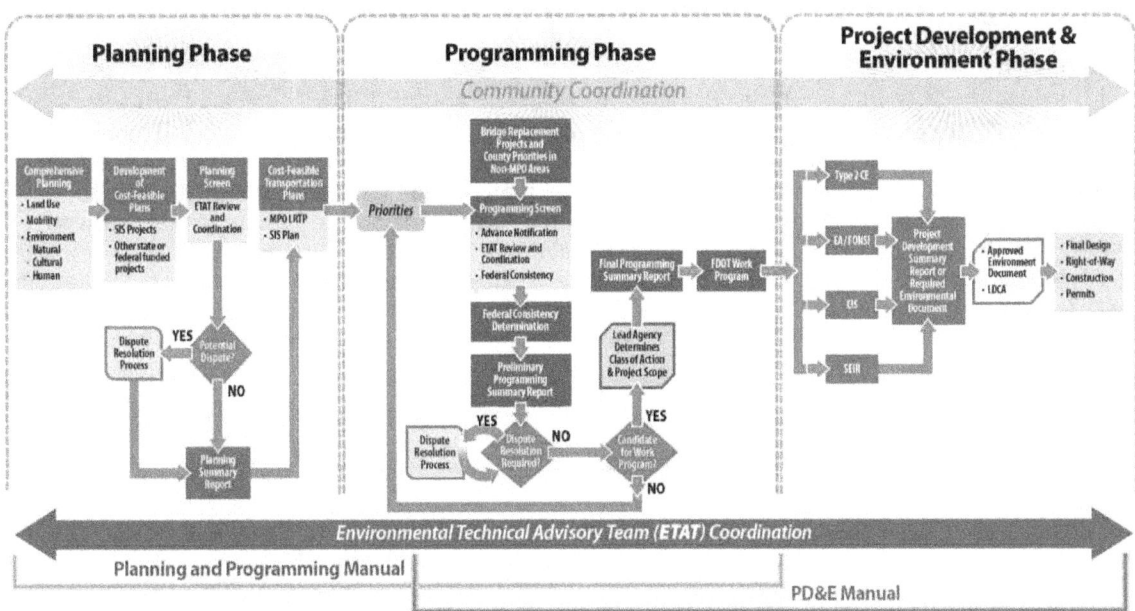

Figure 2. ETDM Process (courtesy of FDOT).

Planning Screen

Comments collected from ETAT during the planning screen assist FDOT and MPOs in determining the feasibility of proposed projects for inclusion in their long-range transportation plans (LRTPs). Reviews and comments at this stage are cursory and provide the project sponsor with a general understanding of the project's purpose and need and affected environment (e.g., the presence of noise vibration sensitive facilities). The project sponsor uses this feedback to acquire a better sense of the proposed project's viability and any critical flaws with the project or a particular alternative. FDOT then revises project concepts to reflect feedback and cost feasibility concerns. Early avoidance or minimization efforts may seek, for example, route modifications that minimize the need for noise walls. Projects then await prioritization by FDOT and MPO management before moving into the programming phase.

Programming Screen

During the Programming screen, projects are under consideration for inclusion in the FDOT Work Program. The Programming screen builds upon the information produced during the Planning Phase by identifying potential avoidance, minimization, and mitigation opportunities and refining or potentially eliminating alternatives, as applicable. The Programming screen initiates the National Environmental Policy Act (NEPA) project scoping process for projects with Federal funding/action. The Programming screen specifically involves soliciting more detailed feedback from agencies and partners in order to determine the necessary environmental document (e.g., Environmental Impact Statement (EIS)) and associated technical studies and permits. FDOT can then use this information to estimate project costs and the project timelines. During this phase, FDOT considers noise in terms of the community's perception and identifies the likely impact on noise-sensitive receptors located in the project area. Noise assessments look at the specific character and environmental context of the project and not just on traffic volume.

PD&E Screen

Finally, the PD&E phase involves using the information generated during the earlier programming screen to prepare the necessary studies and reports needed to accompany the corresponding environmental document (EIS, Environmental Assessment, etc.) that must be developed for the project. At this stage, the range of project alternatives has been defined, making detailed noise analysis appropriate. Also, during this stage, the ETAT will provide technical assistance upon request by FDOT.

Using the EST

Data in the EST are organized by issues such as farmland or noise. A number of mapping tools and imagery datasets are available to support the analysis process. This includes common mapping application tools such as queries and buffers as well as more unique functions such as Google's Streetview (see Figure 3) and an FDOT video log of State-owned roadways. Various aerial imagery datasets are also available, including one-foot resolution digital orthophoto imagery from 2009 or later. FDOT can customize the EST's data to accommodate agency partners' respective reviews. Staff at the GeoPlan Center work with the ETAT to obtain/update existing data layers or to develop new layers. Data-update cycles are layer-dependent, but FDOT works to ensure the most recent dataset is always available.

Figure 3. Google Streetview Integration with the Map Viewer.

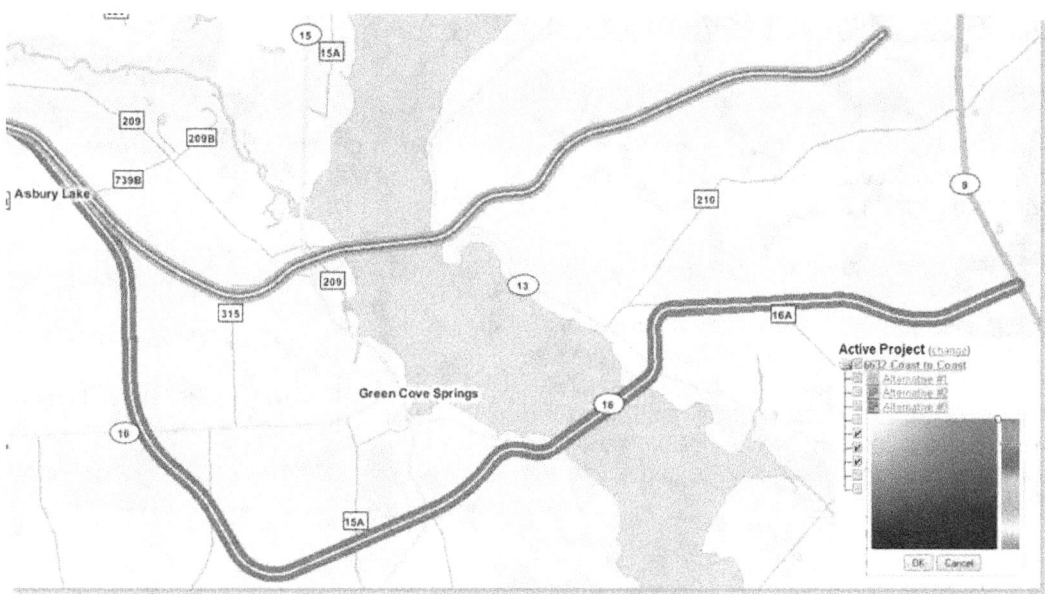

Figure 4. Screenshot of EST.

When resource agencies and partners log in to the EST, they can review a project's purpose and need, description, and associated GIS analysis results. The map interface is customizable and users have the flexibility to determine which issues and project alternatives are visible. Users can also select buffer zones ranging from 100 feet to 5,280 feet (1 mile) to see what data features fall within the specified range of the project.

Overall, considering noise early on in the planning process has led to cost savings and better project outcomes since the potential need for noise walls can be identified before funding is committed to a project. ETDM enables FDOT to resolve or mitigate particular noise issues early on in project development. Identifying a noise issue late can lead to increased project costs and potentially add time to the production delivery schedule.

Future Steps

Because analyzing noise as a stand-alone issue is relatively new, FDOT is still identifying data gaps and potential sources of new noise information to add to the EST.

In the future, FDOT intends to fully update and expand its 2005 noise wall GIS database and has begun to identify the attributes that will be captured. Once the new geodatabase is developed, it will replace the current noise wall GIS database within the EST and also include an automated data reporting system to satisfy information requests related to noise walls. FDOT hopes this noise wall inventory database can be developed in partnership with other FDOT offices that could stand to benefit, such as FDOT's Maintenance Office.

FDOT also noted that they are interested in investigating opportunities to develop future datasets for inclusion in the EST, such as:

- Land uses categorized by noise impact thresholds;
- Noise contours based on topography;
- Traffic data and traffic composition; and
- Ground types, such as soft ground (grass), hard ground (paved areas), and water.

Maryland State Highway Administration (MDSHA)

Background

MDSHA's Office of Planning and Preliminary Engineering (OPPE) and the Office of Highway Development share duties related to the highway noise program in Maryland as the result of a reorganization effort in 2008.

Currently, the Noise Abatement Design and Analysis Team (NADAT) in OPPE is responsible for technical and policy guidance related to noise, review, and approval of all Type I highway noise studies under NEPA, community noise studies, and noise barrier acoustical design (for Type II barriers). There are three staff members within OPPE working on noise issues: the NADAT team leader, an Environmental Planning Division noise specialist, and supplemental consultant support.

> **MDSHA uses GIS in the noise discipline to:**
> - Develop and maintain a noise wall inventory.
> - Format noise-related data for contribution to an enterprise GIS system.

Since its noise program started in 1977, MDSHA has compiled numerous noise data that were in multiple formats, both paper and electronic. These data included noise complaints, community noise impact studies, community highway and development histories/build dates (for Type II analysis), NEPA analyses, after-studies of barrier effectiveness, and inventories of completed Type I and Type II barriers.

Before moving to GIS, most noise-related archiving activities utilized ProjectWise.[15] Over time, MDSHA identified a need for a centralized repository for disparate information that could be accessed by leadership and users in multiple departments. Until recently, however, there was no strong GIS platform available to which information could be contributed. Recently MDSHA developed an enterprise GIS web application called MD iMap,[16] which consolidates all information the agency uses in its transportation programs and decision-making.

NADAT has started preparing and contributing noise-related information, including a noise wall inventory, to MD iMap by focusing on the extraction and organization of data elements related to NEPA noise studies, archiving old studies and reports, and compiling independently-developed databases and inventories. This effort is expected to provide a cohesive and complete information system for the highway noise program in Maryland.

Noise Wall Inventory

Concurrent with its efforts to prepare and contribute noise data to MD iMap, MDSHA hired a consultant to compile an inventory of every noise barrier and its location data in the State, as well as to take multiple photographs of each barrier. The consultant also "field verified" data found in existing records and collected global positional system (GPS) coordinate data at critical locations along the barrier alignment, as well as at locations of design elements such as fire hose connections or maintenance doors. During this process, the consultant incorporated and field-verified other data on the barriers, such as materials used, surface textures and finishes, physical condition/damage, and physical dimensions (wall length and height).

A large portion of these data had been compiled and maintained over the course of the highway noise program, but they were stored in disparate formats and locations. The 2010 effort to compile a comprehensive inventory was able to incorporate many of these older data. For example, the MDSHA

---

[15] ProjectWise is engineering project management software program that facilitates the sharing and review of project engineering information in a single platform.

[16] iMap is available at www.imap.maryland.gov/portal/. iMap is a Flex Application with external systems such as ArcGIS Server 10 and eGIS Web Application. The eGIS system allows various MDSHA departments to add content via the GIS Web Application, build tools and widgets, or build stand-alone applications that they can embed in websites or launch from websites.

Office of Structures developed an inventory of all "small structures," which gathered information on retaining walls, headwalls, culverts, and noise barriers within a searchable database. Georeferencing was not part of this original effort, though each item was assigned a structure number. These data were ultimately combined and incorporated into the noise wall inventory, with the structure numbers serving as the common data element. Information was verified with reference to a PDF archive of past as-built plans for projects with noise barriers.

Once completed, the inventory went through a review process to check for redundancies with other datasets and for quality control purposes. MDSHA aims to complete verification of the georeferencing and proceed with integration of the inventory into MD iMap soon. In the meantime, MDSHA has increasingly relied on Google Maps—particularly its aerial information—when responding to noise-related inquiries from the public or other MDSHA departments, district offices, or local government agencies.

To date, the main benefit of the inventory has been providing inspection and maintenance tracking and a data source for program-level reporting to FHWA and within MDSHA. The completed inventory has also been useful when responding to inquiries from the public, elected officials, and resource agencies.

The total cost of compiling and updating the noise wall inventory has likely been between $100,000 and $200,000, though MDSHA has not estimated the accumulated total costs over the years.

Separate from the noise wall inventory, MDSHA has also maintained a noise complaint inventory comprised of State grid maps with colored dots (a "manual GIS") to define where complaints had been received, letters written, and studies performed. Some complaint data has been converted into electronic databases but has not been georeferenced. There is a less urgent need for regular access to geocoded noise complaints, especially given other available tools such as Google Maps. This is also because all Type II projects identified in the original candidate list from the late 1970s have been constructed and a finite number of potential future Type II projects exist based on MDSHA and FHWA eligibility criteria. The complaints MDSHA receives now are somewhat more irregular, and less likely to involve extended coordination (i.e., not from the same neighborhood). In the past, prior to building the complaint inventory, MDSHA would reference noise complaint maps to see if complaints had been received; multiple individual or repeated inquiries often signified that an area was a strong candidate for a Type II noise barrier.

Future Steps

When all existing noise walls are fully geocoded and integrated into MD iMap, MDSHA envisions expanding the inventory to include proposed walls. GIS would support this effort by outlining geographic relationships between noise receptors, impacted communities, project boundaries, and environmental features. MDSHA is also considering incorporating environmental noise data about planned projects into MD iMap to keep a record of projects and to support reevaluations in the future. This would help provide a useful reference when potential projects are considered or in some cases reactivated after many years.

In the future, MDSHA may also link the inventory to a photolog application used elsewhere in MDSHA called "Visidata." If linked with the inventory, Visidata would allow users to view existing pictures of noise walls or communities. MDSHA also has a video log that could potentially be linked to the noise barrier inventory in MD iMap.

MDSHA does not currently use GIS as part of its TNM noise modeling process but may do so in the future.

Lessons Learned

MDSHA reported the following lessons learned:

- **Be flexible as applications evolve over time.** MDSHA recognizes that its noise barrier data are primarily static information, but that the noise wall inventory could be a "stepping stone" toward developing a fully-dynamic GIS tool. The agency initiated an effort to conduct electronic report archiving and cataloging that could ultimately be geographically linked to noise barriers in an online web application. MDSHA would also like to start collecting information about completed projects through the Highway Performance Monitoring System in order to keep the inventory up-to-date, and to aid in expanding record-keeping to include NEPA-generated data.

- **Ensure that GIS for noise analysis applications are compatible with existing GIS databases.** MDSHA found there was a high level of planning necessary to collect data to add to the noise barrier inventory and migrate it to a GIS-based system. The inventory evolved over many years, and the database was still evolving as staff members began to develop a GIS platform. Developing appropriate database relationships and the applicable platform to integrate all the data and intended applications were challenges. Also, envisioning the full potential use of the inventory and the most appropriate technical design and requirements for the database were difficult. It was helpful to have an existing basemap (MD iMap) at the agency-wide level to which data could be added. Further, MDSHA leadership has been supportive of integrating data into a visual format.

North Carolina Department of Transportation (NCDOT)

Background

NCDOT recently overhauled and updated its traffic noise abatement and modeling policies in accordance with 23 CFR 772. This overhaul included adopting GIS as a regular tool to support noise analyses in responding to significantly shorter project cycles.[17] Prior to 2010, NCDOT's noise assessments generally consisted of using simple distance and source-emission equations and flat and level models, which NCDOT notes sometimes under-assessed noise abatement criteria (NAC)[18] impacts. The only GIS data used to support noise analysis were parcels and existing local roadways. With GIS capabilities now in place, NCDOT considers its noise program to be more formalized and proficient, as well as more beneficial to the public in terms of technical assessment and cost-effectiveness.

> **NCDOT uses GIS for noise analysis to support:**
> - TNM modeling;
> - Design of noise walls; and
> - Assessment of traffic noise levels.

NCDOT's progress in using GIS has been aided by several statewide initiatives and NCDOT's own leadership. Agencies in North Carolina have access to statewide LiDAR data, with 12-inch to 18-inch resolution, as well as a central repository for statewide GIS data collected from State and local agencies. NCDOT leadership has advanced the noise program through its stated recognition that noise walls should not be designed as part of the roadway, but rather as part of the community. NCDOT leadership also worked to ensure that contracts with consultants included provisions that required the submittal of all data collected for a project, rather than just for analyses or reports, which has helped expand the agency's data library.

NCDOT's traffic noise analyses, abatement design, and public outreach are performed by four in-house staff members and several on-call consulting firms. The noise team works independently of NCDOT's GIS staff but relies on the GIS expertise when available, since virtually all of NCDOT's noise activities now utilize GIS. NCDOT is officially not allowed to use Google or Bing mapping tools in its work, as it does not hold an enterprise license.

NCDOT does not yet have a comprehensive noise wall inventory of existing walls, but data on newly constructed walls are shared with FHWA. There are approximately 50 miles of walls and earthen berm sound barriers throughout the State.

GIS and Noise Modeling

To conduct its noise modeling, NCDOT's noise staff transforms GIS data into a format (usually a spreadsheet) compatible for uploading to TNM. GIS data helps streamline the TNM process since noise staff members may not have time to complete an on-the-ground survey of the project area. NCDOT generally conducts a preliminary noise assessment by viewing GIS features (such as roadway composition and topography) to identify the potential exposure of noise-sensitive land uses to traffic noise. Using TNM, NCDOT then performs a detailed design analysis for each project alignment by modeling every receptor. All receptors are linked to an address, owner name, and can be replicated for future analysis. In the past, cursory screening analyses of projects sometimes led to incorrect abatement modeling outputs. NCDOT now justifies detailed analysis even if construction does not proceed because the cost of doing so ($5-10,000) is much lower than building unnecessary noise walls ($1,000,000 -

---

[17] *"NCDOT Traffic Noise Abatement Policy"* is available at
https://connect.ncdot.gov/resources/Environmental/Compliance%20Guides%20and%20Procedures/2011%20NCDOT%20Traffic%20Noise%20Abatement%20Policy.pdf. The accompanying *Traffic Noise Analysis and Abatement Manual* is available at
https://connect.ncdot.gov/resources/Environmental/Compliance%20Guides%20and%20Procedures/NCDOT%20Traffic%20Noise%20Analysis%20and%20Abatement%20Manual.pdf.
[18] For a definition of NAC, *see* www.fhwa.dot.gov/environment/noise/regulations_and_guidance/faq_nois.cfm#note17.

$2,000,000 per linear mile). Furthermore, such analysis can be used in the event a project alternative is revived or modified at a later date.

NCDOT maintains a rigid tolerance for noise modeling, which allows the agency to produce assessments at a very high level of detail. While +/- 3 decibels is the FHWA-recommended minimum and "industry standard," NCDOT applies a tolerance of +/- 1.7 decibels. NCDOT notes that assumptions built into Reference Energy Mean Emission Level (REMEL) data[19] included in TNM are not as accurate as they perhaps could be, but TNM model(s) can be refined on a project-by-project basis to attain a leaner level of tolerance. NCDOT's goal is to utilize the TNM model in a way that shifts focus from noise sources to noise receivers. GIS advances this effort by enabling the inclusion of more detailed data related to a project's environmental context.

GIS has also enabled NCDOT to move forward on noise modeling and design noise walls without waiting for project survey data. For example, 11 noise walls were designed for a maintenance and widening project on Interstate 40 in Raleigh. Detailed analysis of surrounding neighborhoods indicated that 23,000 linear feet of previously unanticipated noise wall—at a cost of $12 million for 11 walls total—were necessary to respond to hundreds of impacts to the more than 1,500 residences in the vicinity of the project. GIS data from Wake County were used to model these neighborhoods in TNM, which involved validating 30-40 discrete monitoring locations (see Figures 5, 6, and 7). Even without survey or preliminary design data, NCDOT's noise consultant was able to design the noise walls.

Figure 5. "Hand Modeling" of I-40 in TNM Using GIS.

---

[19] REMEL data are included in noise prediction models such as TNM. REMELs represent the "maximum, energy-averaged, A-weighted sound level of a vehicle type." (Wayson, Ogle, and Lindeman, 1994: www.adc40.org/docs/paper_award/1994%20Paper%20Award.pdf).

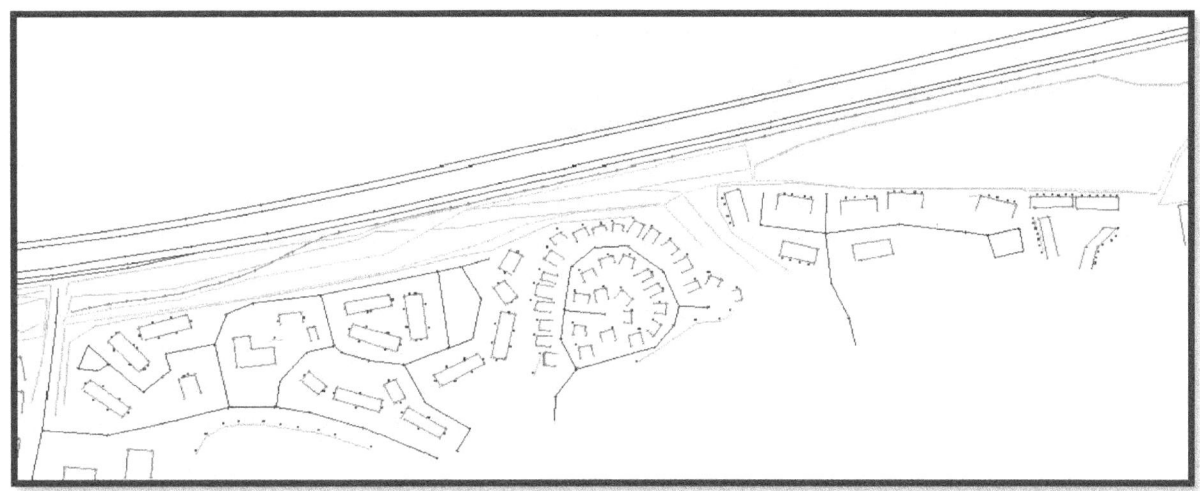

Figure 6. Resulting Existing Condition TNM Model for I-40 Widening.

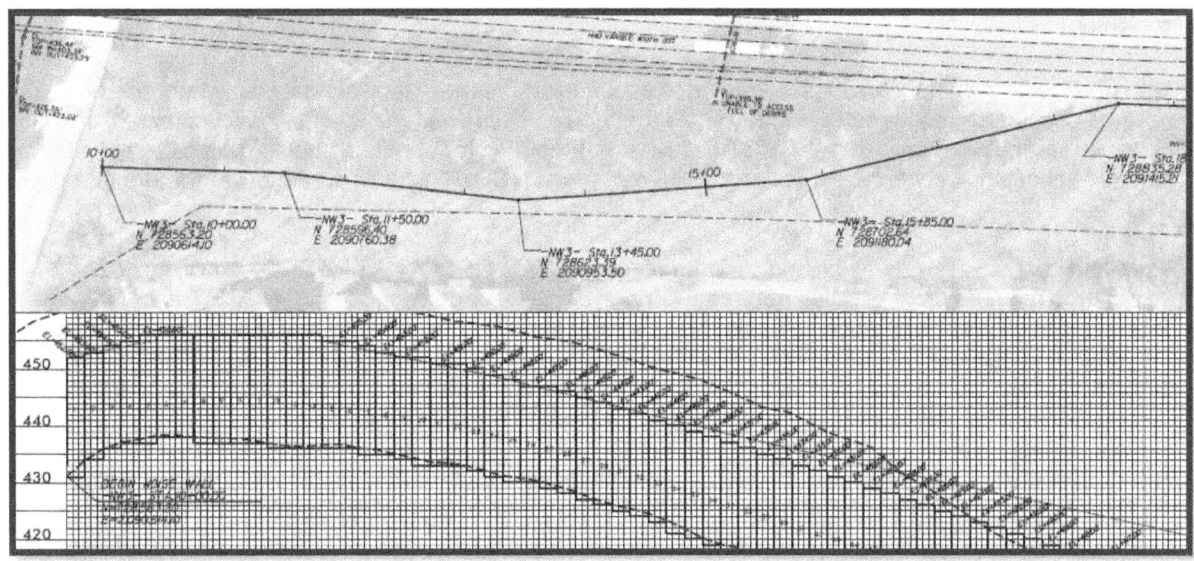

Figure 7. Resulting Noise Wall Design for I-40 Widening.

Data

NCDOT uses the same baseline GIS data each time it runs TNM, including ground elevation or elevation contours, receptor locations, vegetation, land cover, drainage, and utilities. NCDOT also adds additional "acoustically relevant" data as available/appropriate, such as existing roadways, bodies of water, large infrastructure, and tree zones (tree zones are subjective and are only used to validate the noise model).

The most centralized GIS data source for NCDOT's noise assessments and modeling is a statewide repository called NC OneMap.[20] The State of North Carolina began developing NC OneMap in 2003. It is a free, online, and publicly-accessible database containing data from numerous State and local agencies (see Figure 8). Information in the database includes wildlife, wetlands, land uses, and other topic areas. Currently, a minimal amount of county-level data available exists in this resource, but eventually NC OneMap is expected to become the single repository for local and statewide GIS data. NC OneMap data that is most useful for NCDOT's noise team includes information on elevation contours, topography, vegetation, hydrography, utilities, and building construction elevations.

---

[20] NC OneMap is available at www.nconemap.com/.

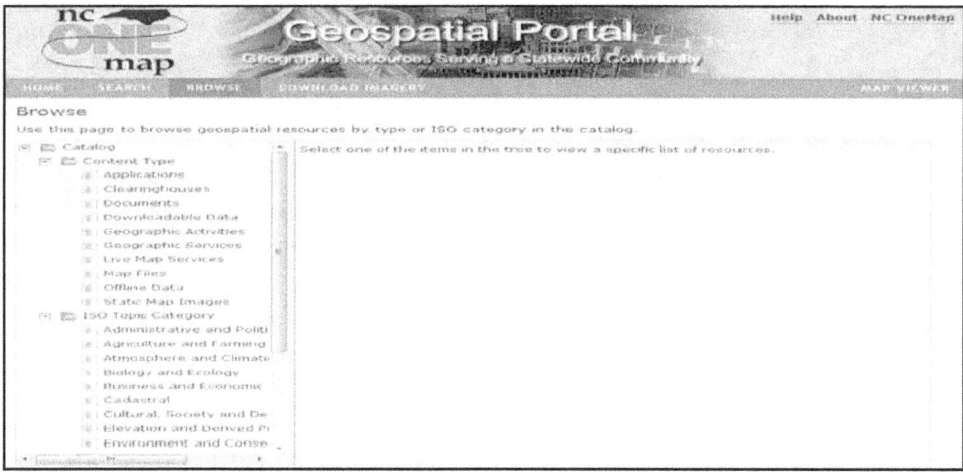

Figure 8. Screenshot of NC OneMap's Geospatial Portal (data browse function).

Benefits

One of NCDOT's primary lessons learned is to avoid using the term "spatial analysis" when referring to noise assessments, as these terms may imply that noise assessments are simply calculations of distance between sources and receivers. NCDOT believes that the physics of sound should be factored into the spatial relationship between sources and receivers, which NCDOT finds is made easier through the use of GIS.

Time and cost savings are two noted benefits of using GIS in NCDOT's noise analysis. While GIS sometimes requires more detailed front-end work, especially for noise modeling, NCDOT found that the total amount of work involved in noise analysis has decreased. If accurate data are added to TNM from the beginning, the need to make large reconfigurations of noise mitigation plans as a result of minor changes to a project have been minimized. Furthermore, due to limited staff resources, NCDOT does not believe it would be able to meet project deadlines without the help of GIS.

NCDOT has found that GIS data are often more detailed and more useful than survey data collected for a highway project. For instance, the ground line for the terminus of a noise wall was off by 1.5 feet on one project because survey data used 50' surface points. NCDOT notes that GIS data would have been more accurate in that circumstance.

NCDOT has been able to design noise walls and berms using only GIS data because the determining factor in designing a noise wall is an elevated line in the sky, called the "top of the wall" (or acoustic) profile. This is the line the top of the wall must take in order to mitigate noise as it travels from source to receiver. The noise wall can be designed using GIS before a survey is complete because as long as the top of the wall is accurate, the ground line can later be adjusted as needed. Since GIS data may not pick up on every dip or rise of the ground, using a top of the wall profile allows for greater design flexibility to prevent wall height from undulating across different elevations. For this reason, NCDOT no longer specifies wall heights.

The top of the wall profile is generally tied to the roadway vertical profile, though it may also be tied to the designed offset (such as the centerline) of the roadway. The wall's final horizontal alignment and ground profile (i.e., where the wall will meet the ground) is ultimately confirmed through a field survey. If the survey indicates a significant change in the roadway's designed offset or horizontal or vertical alignment, the top of wall profile may have to be adjusted, depending on its adjacency to the edge of pavement. If the top of wall profile is designed with a lesser correlation to the edge of pavement (such as by selecting a location at the top of a cut-section or near the right of way limits, for example), then only very significant changes in the roadway's horizontal and/or vertical alignment will necessitate a revision of the top of wall profile.

GIS also assists NCDOT in identifying details on receivers (such as building shape, height, presence of berms, etc.), which NCDOT believes is sometimes an undervalued aspect of noise analysis. This may be because roadway engineers are not typically trained to collect information beyond the highway right of way, where receivers are located. Based on project experience, NCDOT believes that appropriately detailed noise models result in more accurate traffic noise level and impact prediction, as well as more efficient and aesthetically pleasing abatement designs. Furthermore, detailed modeling outside the roadway also benefits any future road widening projects, since updating the noise model only requires a change in the modeling of proposed roadway design features.

Challenges

While there are numerous benefits associated with use of GIS, NCDOT reported that it can be difficult to rely on NC OneMap as its main source of GIS data. The map can be unstable and sometimes crashes. Some staff members have found the system's navigation to be cumbersome at times. Additionally, NC OneMap does not currently have many detailed county-level datasets that are important to NCDOT's noise analyses. Collecting data from individual local governments can be time-consuming, and this information sometimes contains inconsistencies due to lack of coordinated data collection at the local level. NCDOT notes that, for some counties, it is particularly difficult and costly to find quality elevation data.

GIS data from areas outside the highway right of way are included in TNM, though to use this information, assumptions must sometimes be made about the data's accuracy. Per FHWA regulations, noise impacts cannot be assessed using only noise contours created by GIS; only TNM and the algorithms it contains must be used.

It can be difficult to validate noise models for new roadways that have no existing noise sources since there are no "real world" data to confirm a model's predictions. Approximately, three-quarters of NCDOT's projects do have existing noise sources, so these challenges are not typical. When applicable, NCDOT will collect ambient noise level data at appropriate locations, making iterative adjustments to TNM to ensure that the model's predicted noise levels are within acceptable tolerances of those monitored in the field.

Future Steps

NCDOT's noise team is currently building a statewide library of all TNM models and is considering developing a GIS repository for all noise analysis data, results, design, and recommendations for use in future projects or to respond to public requests and concerns. Used in this way, NCDOT considers GIS a "forward-working" tool rather than simply a storage system. NCDOT also sees potential in using such a system to link the date, manufacturer, and condition of noise walls with environmental data so as to more proactively track and predict maintenance needs for the future.

NCDOT believes that developing an agency-wide GIS-based tool might enhance the noise screening process if criteria such as elevation, forestation/vegetation, structures, water features, roadways, and buildings were included, as well as existing noise abatement measures. This type of tool could also contain a digitized catalogue of noise abatement measures. However, NCDOT currently does not have staff capacity in terms of skills or availability to build such a tool. NCDOT might consider building this type of tool in the future once its consultants are fully trained on NCDOT's and FHWA's new noise abatement policies and procedures. NCDOT also plans to hire consultants to continue to educate the noise team on GIS techniques, since NCDOT's other GIS staff are already working at capacity.

Finally, NCDOT would ideally like to compile a digitized noise wall inventory to add to NC OneMap and make it available to local governments.

Ohio Department of Transportation (ODOT)

Background

ODOT's noise section is in the Office of Environmental Services within the Planning Division.[21] One staff member in the central office is responsible for a majority of ODOT's noise analysis.

ODOT, which has about 181 miles of noise walls across the State, uses GIS to locate and catalog noise barrier walls. Geospatial data from Google Earth and Bing Maps are also used to support traffic noise modeling analysis.

> **ODOT uses GIS and geospatial data to:**
> - **Locate noise barriers.**
> - **Support traffic noise modeling analysis.**
> - **Build institutional knowledge of noise walls.**

ODOT identified a need to capture information about noise barriers across the State in a spreadsheet (see Figure 9). Originally in Quattro, the spreadsheet was converted to Excel in 2004-2005 when the State switched from a Corel Office Suite to Microsoft Office. The spreadsheet was later enhanced to help staff better and more consistently address questions from ODOT District Offices, the public, and others about the noise barriers and their placement.

In developing the spreadsheet, ODOT considered what type of information would be needed to satisfy public information requests. When available, project plans were used to capture a range of information about the barriers, including their location, when they were constructed, their square footage, materials used in construction, length, and total costs (based on an average of $25/square foot).

ODOTDIST	COUNTY	ROUTE	ROUTE #	ROUTE DIR	RTE_SLM	PID	CJN	Year	CITY	METROAREA	COMMENTS
DISTRICT 4	STA	IR	77	S	9.70	20412	583(05)	2005	CANTON	Canton, OH	77 SB from Norfolk Southern Railroad to 6th St
DISTRICT 4	STA	IR	77	S	10.30	83373	243(09)	2009	CANTON	Canton, OH	77 SB from W Tuscarawas St to Dueber Ave
DISTRICT 4	STA	IR	77	N	10.50	20412	583(05)	2005	CANTON	Canton, OH	NE quad of 77 and SR 172 Tuscarawas St
DISTRICT 4	STA	IR	77	N	10.60	83373	243(09)	2009	CANTON	Canton, OH	SE quad of 77 and 4th St
DISTRICT 4	STA	IR	77	S	10.60	20412	583(05)	2005	CANTON	Canton, OH	77 SB from 4th St to SR 172 Tuscarawas St
DISTRICT 4	STA	IR	77	S	10.70	20412	583(05)	2005	CANTON	Canton, OH	SW quad of 77 and 4th St
DISTRICT 4	STA	IR	77	S	10.70	20412	583(05)	2005	CANTON	Canton, OH	77 SB from 10th St to 4th St
DISTRICT 4	STA	IR	77	S	11.10	20412	583(05)	2005	CANTON	Canton, OH	77 SB north of 10th St along Harrison Ave Connector
DISTRICT 4	STA	IR	77	S	11.20	20412	583(05)	2005	CANTON	Canton, OH	SW quad of 77 and 13th St
DISTRICT 4	STA	IR	77	S	12.60	21610		2004	CANTON	Canton, OH	77 SB north of Croydon Dr NW and US 62
DISTRICT 4	STA	IR	77	S	12.70	21610		2004	CANTON	Canton, OH	77 SB from Orchard Park Dr to Croydon Dr
DISTRICT 4	STA	IR	77	S	12.80	16376		2000	CANTON	Canton, OH	77 SB on Orchard Park Dr overpass
DISTRICT 4	STA	IR	77	S	12.90	16376		2000	CANTON	Canton, OH	77 SB north of Orchard Park Dr overpass
DISTRICT 4	STA	IR	77	N	13.20	16376		2000	CANTON	Canton, OH	SE quad of 77 and 38th St
DISTRICT 4	STA	IR	77	S	13.40	16376		2000	CANTON	Canton, OH	SW quad of 77 and 38th St
DISTRICT 4	STA	IR	77	N	13.50	16376		2000	CANTON	Canton, OH	77 NB 38th St Overpass
DISTRICT 4	STA	IR	77	S	13.50	16376		2000	CANTON	Canton, OH	77 SB 38th St Overpass

Figure 9. Screenshot of Select Fields from ODOT's Noise Barrier Spreadsheet.

Developing and updating the spreadsheet was a three-phased effort. The first phase involved working with a consultant to conduct a limited inventory of a few central Ohio counties to develop a proof of concept, followed by a second phase focused on collecting a statewide inventory. In 2007, ODOT hired the same consultant to add newly-constructed barriers to the inventory and update the viewer application. Additional data were added including panel dimensions, wall elevation and height, and GPS coordinates. The consultant also collected information on any points of damage to better assess the condition of the barriers. Additionally, the consultant took pictures of damaged areas along the walls as well as general pictures of the front and back of each wall. Collecting these data took six to eight months. Information on

---

[21] The noise section's website is available at
www.dot.state.oh.us/Divisions/Planning/Environment/NEPA_policy_issues/NOISE/Pages/default.aspx.

wall contractors and material manufacturers were also compiled by communicating with district offices and suppliers.

There were some challenges in this effort. For instance, data was delivered in shapefile format, though at the time ODOT used GeoMedia (ODOT is the only State agency in Ohio that does not use ArcGIS.)

After the field data collection effort was completed, the consultant created a user-friendly ESRI-based desktop viewer to store data from the spreadsheet. ODOT aimed to provide each of its districts with access to the desktop application as a way to promote data-sharing across districts. However, over the course of developing the viewer, ODOT's IT department updated its operations structure. New IT policy prevented the desktop viewer from accessing multiple servers, and challenges were encountered in sharing data across firewalls and obtaining administrator rights to install the viewer. Additionally, some districts used different geospatial software, making it difficult for users to consistently access or update data. To date, these challenges have not been resolved. As a result, ODOT has moved away from developing the ESRI viewer and focused more on using the spreadsheet as a way to store noise wall information.

When the reporting requirements of 23 CFR 772 were established, ODOT staff determined that it could use the spreadsheet to respond more easily to these requirements. Staff decided to georeference a portion of information in the spreadsheet (e.g., geographic coordinates of Type I and II noise walls built since 2005) using GeoMedia. This process took approximately three months.

ODOT has georeferenced all of the information from the noise barrier spreadsheet into GeoMedia. Staff members concurrently use the spreadsheet and GeoMedia to respond to 23 CFR 772 reporting requirements.

The GeoMedia tool has evolved over time. Initially, it functioned like the spreadsheet as a basic repository of information, helping ODOT address questions such as the age of the noise barriers, their total cost and square footage, what material they are comprised of, when they need to be replaced, and the length or number of noise barriers within a particular ODOT district. While the tool continues to be an information storehouse, it has also become a management tool that provides a better way for ODOT to manage maintenance or replacement schedules as well as other needs for the noise barriers. Data in the GeoMedia tool and spreadsheet are updated on an as-needed basis.

Most recently, ODOT staff has digitized information on any replacement walls constructed since 2005, which helps staff members know the lifespan of a noise wall. The tool now contains information on all noise walls constructed since 2005 (information was gathered from project plans and field surveys).

Total cost to conduct the pilot data collection for the spreadsheet, complete the full statewide data collection, develop the GeoMedia viewer application, and update the noise wall inventory was approximately $160-170,000.

Geospatial Data for TNM Analysis

ODOT is also using geospatial information from Bing Maps and Google Earth to support traffic noise modeling analysis. Within the past year, staff have focused on gathering information from Google Earth using software called Maptool, available from Zonum Solutions[22] (see Figure 10). Maptool is essentially a pre-programmed tool that manipulates Google Earth to find data on elevations, distances, and areas. It also allows ODOT staff to easily collect data on receptors such as homes, churches, schools, and parks.

---

[22] Zonum Solutions contains a variety of free software tools. More information is available at www.zonums.com.

Figure 10. Screenshot of Maptool.

Maptool is used to gather all geospatial points from Google Earth to input into TNM to predict future noise levels for specific proposed projects. Using TNM, staff then assess potential noise barriers for transportation project sites and report on whether constructing a noise barrier is reasonable and feasible.

ODOT also uses data from web-based tools to support its noise analysis activities. ODOT uses Google Earth data to depict the locations and shape of potential barrier designs for proposed road projects. The department also uses data from Bing Maps to support noise barrier analysis. These data are supplemented by ODOT's statewide LiDAR data.

Benefits

ODOT has found benefits in using Google Earth data to support traffic noise modeling analysis. For example, roadway design files[23] are a source for data to input into TNM, but these files typically do not include receiver (e.g., homes, schools, hospitals) or noise wall locations. In the past, when using roadway design files for input to TNM, ODOT staff members often had to estimate receiver and noise wall locations, which led to some data inconsistencies. Georeferencing receiver locations found in LiDAR data also turned out to be too cumbersome and time-consuming. Google Earth contains georeferenced information about both roadways and receiver locations; as a result, staff can now look to one information source, saving time and supporting data accuracy. One of the most significant benefits of using Google Earth is that its LiDAR data for Ohio comes directly from ODOT, as the agency recently (in the last year or two) donated its data to Google Earth. As a result, staff have the reassurance of knowing that Google Earth LiDAR data for Ohio are as accurate as possible.

ODOT has not developed any formal performance measures to evaluate or assess the GeoMedia tool or use of Google Earth to support traffic noise modeling analysis. On the basis of anecdotal information, the agency believes that the inventory and Google Earth data "meets or exceeds" its needs. In particular, staff noted that use of Google Earth data has streamlined noise analysis and has led to significant time-savings. Additionally, the GeoMedia tool has helped ODOT obtain information more quickly and easily. It has provided both cost and time-savings in responding to questions from district offices and the public about the noise walls.

---

[23] Design files are electronic packages of information about a constructed roadway.

Challenges

While using Google Earth data was advantageous, there were some difficulties. For example, elevation data from Google Earth was not always accurate. In some cases, when obtaining elevation information for a noise wall located on an overpass, Google Earth would frequently pinpoint coordinates for the road under the overpass without "recognizing" the existence of the bridge. In these cases, ODOT staff needed to develop workarounds.

ODOT also recognizes that not all of the information in its noise barrier inventory spreadsheet is accurate, as some data pulled from project plans have not been field verified. Staff members are continually conducting "detective work" to make the information as accurate as possible.

Furthermore, staff recognized that ODOT leadership currently considers GIS as a support tool, which increases the standards in terms of making the case for further investment in GIS tool development.

Future Steps

In the future, ODOT would like to measure the individual panels that form noise walls and add this information to both the spreadsheet and GeoMedia tool. This information would help ODOT staff members better understand when the panels need to be replaced. Panel information would need to be collected in the field, as data on panel dimensions are not captured in project plans.

The agency might also move forward to digitize all of the information contained in the noise barrier spreadsheet, including data from walls built before 2005, to provide a comprehensive picture of all noise walls in the State. However, this is not currently considered to be a critical need for the agency.

Lessons Learned

ODOT noted the following lessons learned:

- **Coordinate with IT staff.** It important to communicate with IT staff as early as possible when building GIS applications to ensure that all staff are "on the same page" in regards to technical requirements and standards.

- **Consider new uses for existing applications.** ODOT's spreadsheet evolved from a simple repository of information to a mechanism that would help staff respond to 23 CFR 772 reporting requirements. The way in which tools are used can change over time; agencies should be creative and open to considering potential new uses.

- **Consider online tools to support noise analysis.** After conducting a scan of online tools, ODOT identified Maptool as a good solution. Maptool is easy to use, free, and has significantly helped to streamline the process of assessing noise impacts. However, agencies considering use of Maptool or other similar tools should verify that the information they contain (particularly elevation data) is accurate and has been properly validated or documented.

Tennessee Department of Transportation (TDOT)

Background

TDOT's Air and Noise Analysis Section is composed of one primary staff person. The section is part of TDOT's Environmental Division.[24]

Between 2003 and 2005, TDOT began a Type II program to facilitate the construction of "retrofit" noise walls in older neighborhoods impacted by traffic noise from existing highways. As a first step toward starting this program, TDOT's Commissioner authorized a comprehensive Type II study to identify potential locations of Type II noise barriers along the State highway system. In response, TDOT developed an informal, internal GIS database cataloging information on existing noise walls (Type I walls) as well as potential locations of retrofit barriers across the State.

Currently, the primary users of the database are TDOT's environmental division staff as well as other TDOT personnel that oversee materials and maintenance. TDOT is exploring whether the database could be used to help communicate to the public about noise walls, but an approach has not yet been determined.

> **TDOT uses GIS for noise analysis to:**
> - **Catalog information about Type I and Type II barriers.**
> - **Support noise modeling.**
>
> **In the future, TDOT may use GIS for noise analysis to:**
> - **Geocode noise-related complaints.**
> - **Communicate with the public about noise walls.**

GIS Noise Wall Database

TDOT's GIS noise wall database uses an ArcGIS platform and contains an array of information on the State's existing Type I noise barriers as well as potential Type II noise barrier locations.

Type I noise barrier information in the database includes location, construction year, and construction materials. Type II information includes the locations of residential communities along the interstate highway system, estimates of existing sound levels, metadata, and locations that do and do not qualify for Type II barriers, as well as the reasons for ineligibility.

The database has been updated several times since the first version was completed in 2005. However, there are some Type I attribute data fields, such as length, height, and construction cost, which have not been populated due to lack of staff time. However, TDOT plans to complete the database in the future.

The initial purpose of the database was to respond to the TDOT Commissioner's request to identify locations for retrofit barriers. The database was used as an initial screening tool for potential Type II-eligible areas. It allowed TDOT to identify potentially qualifying neighborhoods within buffers adjacent to interstates. Each potentially qualifying area was evaluated and TNM modeling and barrier analyses completed to identify "benefitted residences" in each area. Where available, TDOT also used county property assessor data to determine the construction dates of residences in potential Type II areas. Ultimately, TDOT determined that 22 neighborhoods were eligible for Type II noise walls at a total cost of about $30 or $40 million. To date, three Type II noise walls have been constructed.

The database provides a formal way to organize information about noise walls that was not previously documented. For example, Federal funding cannot be used to construct a Type II noise wall for an neighborhood if a barrier was previously determined not to be feasible or reasonable as part of a Type I noise study. TDOT needed a way to quickly identify if a neighborhood had been studied previously for a Type I wall and then deemed ineligible. As a result, all Type I projects on the interstate highway were

---

[24] TDOT's policy on highway traffic noise abatement is available at www.tdot.state.tn.us/environment/airnoise/pdf/TDOTNoisePolicy520-01.pdf.

included as objects in the GIS database, which allowed staff to quickly identify if adjacent neighborhoods were ineligible for Type II walls. TDOT's access to statewide parcel data also facilitated this process.

In the future, staff members hope to use the database to better manage the Type I noise wall inventory and to answer questions about walls' cost, number, and location. TDOT also plans to make some modifications to the GIS database to allow development of queries and reports that would generate specific data required by the Federal noise regulations in 23 CFR 772.

The database continues to evolve. TDOT is currently adding information on noise-related complaints to an Access database. The agency would like to geocode the complaint location, contact information, date, and TDOT's response and add this information to the GIS database to build a comprehensive library of all noise complaints. Beyond helping staff better visualize spatial patterns among complaints, geocoded complaints would allow staff to more easily access detailed information, allowing for quicker and more thorough responses. TDOT is still determining how to proceed with geocoding complaints and identifying potential users of the complaint inventory. There is currently no definite timetable for when this effort will occur.

GIS for Noise Studies, TNM, and Noise Modeling

TDOT is not currently using GIS on a regular basis to conduct noise assessments, although TDOT has used GIS for several individual noise applications.

In one example, TDOT identified potential Type I noise barrier locations for several alternatives using GIS data developed for a major study on the feasibility of building a third bridge across the Mississippi River in Memphis. GIS data on land uses were used to help assess each alternative's potential to create noise impacts. Aerial photography was used for locations where GIS data was not available. TDOT then worked with a consultant to summarize the noise impact and noise barrier information in a report.

TDOT also uses some GIS data with TNM on a project-by-project basis. TDOT developed a manual that outlines the process for using TNM to conduct noise modeling. The manual suggests using GIS "when possible" to support modeling but does not specify when or how to use GIS.

Geospatial data from TNMap Portal, a web-based, statewide GIS system developed by the Tennessee State Office of Information Resources[25] is used to support noise modeling (see Figure 11). TDOT noise staff also use geospatial data from Google and Bing Maps extensively to augment modeling efforts. In particular, the Google street view and Bing "bird's eye" view features can be used to help identify the unique topography of a particular area and to address queries and complaints about noise walls. While this feature is not available for all roads (either in Google or Bing maps), it can be helpful as staff do not have to conduct a site visit to obtain information.

---

[25] TNMap Portal is available at http://tnmap.state.tn.us/portal2/.

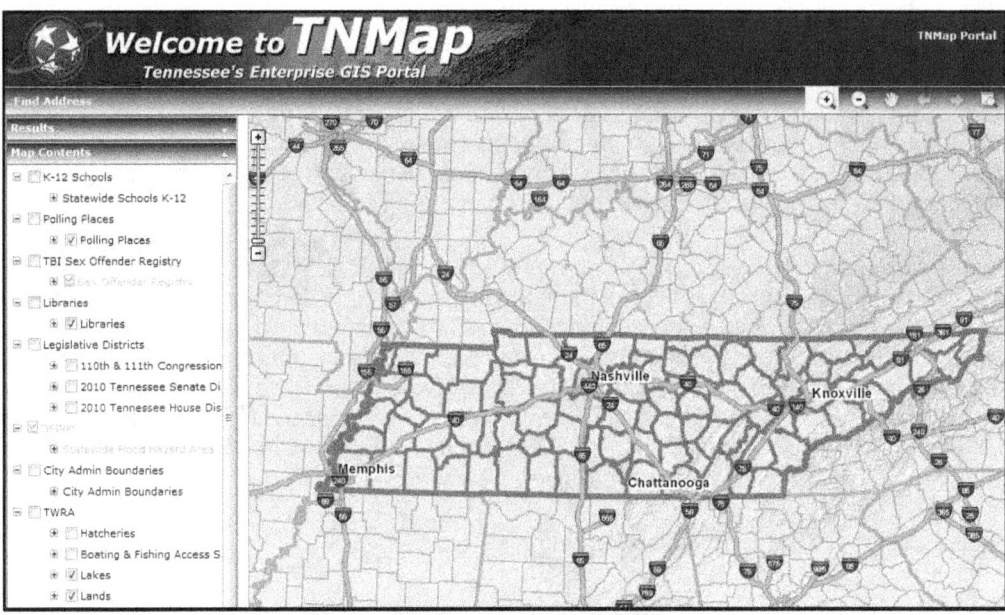

Figure 11. Screenshot of TNMap.

There are no significant costs associated with TDOT's use of GIS to support noise analysis or its development of the GIS database. Some municipalities might charge TDOT for data, though they will typically provide data for free if TDOT documents that the data will be used for project analysis on behalf of the State. In the future, reciprocal agreements might be formalized between local government agencies and TDOT to cover data-sharing.

Benefits

Making GIS information available "at its fingertips" has been very beneficial to TDOT. Documenting past highway program activities in the GIS database has helped inform new policy directions. GIS data have given TDOT a much better understanding of noise impacts and mitigation measures for current and past transportation projects.

Additionally, the quality of county-level GIS data has improved as technology has evolved, making it progressively easier for TDOT to use GIS data to support noise modeling. Higher quality data in noise modeling improves accuracy and decision-making.

Challenges

TDOT has sometimes had difficulty recruiting and retaining staff members with the expertise and training to make best use of GIS capabilities.

Obtaining high-quality GIS data and detailed local-level information for areas beyond the highway right of way has also been a challenge for TDOT, since survey crews generally focus on collecting information only within the right of way. Most roadway design files[26] show only a small area beyond the highway right of way, even though TDOT often needs to look at much larger areas to accurately assess noise impacts.

Furthermore, GIS data is not always available for more rural areas of the State. In some cases, TDOT staff members will use aerial data, property assessors' maps, local-level information, TDOT's project design files, or information from the U.S. Geological Survey to develop more accurate data of a larger area. However, this can lead to challenges with resolution. Currently, TDOT is using a wide variety of

---

[26] Design files are "packets" of geospatial information coded in particular ways. These files can be shared and uploaded into a GIS.

geospatial data sources since it does not have a centralized source that provides all the needed information. Searching for data can add time and cost to noise assessments.

Using GIS in conjunction with TNM also presents challenges as TNM does not currently utilize a GIS interface and is somewhat limited in the extent to which external data can be used. As a result, TDOT personnel must identify "workarounds" that involve translating data into appropriate file formats to use with TNM. Furthermore, TDOT's GIS database is on a statewide scale while TNM outputs are on a project scale, leading to challenges and potential discrepancies.

Future Steps

Ultimately, TDOT envisions adding the GIS database to its external website. This would support TDOT's communication with the public by allowing individuals to access noise wall-related information for specific areas. It would also enable TDOT to respond more efficiently to questions from the public about noise walls and the reasons why neighborhoods either did or did not receive one.

TDOT has already started developing an enterprise GIS-based environmental management system and noise staff would like to see that work continue. Several years ago, TDOT noise staff met with an internal web developer to provide a list of data fields relevant to noise that would be useful to include in this system. The system is envisioned as a centralized clearinghouse of information that would assist noise staff in meeting NEPA requirements (there is no definite delivery date for when the system would be completed; it is an ongoing effort). Using data from this system, for example, noise staff could develop a detailed map of an area during a project's planning phases as a way to assess potential noise impacts. In the meantime, noise staff have looked to a variety of sources, including resource agencies such as the U.S. Fish and Wildlife Service, to conduct this type of screening.

To be most useful to support noise analyses, the system would need to capture elevation and contour data as well as information on land use type and category as defined in 23 CFR 772. The benefit of the application would be in compiling information from a variety of sources into a "one-stop shop."

Lessons Learned

TDOT noted a few lessons learned from its experiences:

- **Plan ahead when developing a GIS tool for noise analysis.** When TDOT developed its GIS noise wall database, it initially focused only on meeting the Commissioner's request to study proposed Type II locations and did not have a broader vision for the tool. Taking time to identify this vision would have helped developers create the database in a more strategic manner.

- **Develop instructions for consultants to ensure consistency in how noise information is collected.** Situations might arise that make it difficult to know how to inventory noise walls. For example, a question of whether a noise wall that ends before a bridge and then continues on the other side should be inventoried as one continuous wall or as two walls might arise.

- **Consider how GIS tools for noise analysis can support public outreach.** GIS data have helped TDOT staff communicate information to the public, particularly through the development of maps or other visualizations that help explain why a community will or will not receive a noise wall.

Virginia Department of Transportation (VDOT)

Background

VDOT's Integrator 2.0 enterprise GIS is an internal, web-only application that has been in use since 1998. Integrator contains approximately 250 layers that are maintained and used by various VDOT business units to support a spectrum of business functions (however, VDOT does not maintain most environmental data in this web application). These layers include a variety of noise-related information, including digital elevation model (DEM) data layers and a sound wall location data layer.

Three VDOT staff members are augmented by two on-call consultants who focus only on noise issues.[27] There are about 130 miles of Type I noise walls across the State; VDOT does not have a Type II noise program.

Within the last several months, VDOT's noise staff members initiated an effort to update Integrator's sound barrier data layers using data obtained over the past five years from Fugro Roadware, an infrastructure data collection company. The update will help VDOT respond to the reporting requirements of 23 CFR 772 and will support VDOT in more accurately documenting the locations of sound walls.

VDOT uses GIS for noise analysis to:
- **Respond to 23 CFR 772 reporting requirements.**
- **Obtain more accurate information on elevations and land features for noise modeling.**
- **Develop noise report graphics.**
- **Aid the evaluation of TNM-predicted sound levels and design of noise barriers.**

The update is expected to be completed within the next one to two years. VDOT's maintenance division is in the process of writing a manual to guide the inventory update and associated data "ground-truthing." There are no specific costs associated with this update other than staff time.

By updating Integrator's sound barrier layers, VDOT aims to help staff better assess the conditions of noise walls, which will then in turn help determine appropriate steps for noise wall maintenance.

Uses of GIS

Integrator's current sound wall layers show the locations of noise walls across Virginia (see Figure 12). This layer was created using 2002 aerial imagery. The imagery did not have a fine degree of resolution and when the information was digitized, data on retaining walls and fences were erroneously included in the layer along with the actual noise walls. As a result, the layer indicated that VDOT had over 450 more miles of noise walls than it actually does. That layer has since been edited to remove many of the erroneous barriers and now totals 123 miles of sound walls; this will soon be replaced by the new layer created from VDOT's current data collection efforts.

As part of this effort, VDOT is currently working with Fugro Roadware to collect a variety of digital pavement data and imagery as well as pavement information on all of the State's primary roadways, ramps, and loops.[28] Fugro Roadware will also collect various data points on noise walls, including information on the walls' start and end points, construction materials, and conditions. Subcontractors will also conduct visual inspections of each mile of imagery. Ultimately, VDOT will digitize this information and update the existing Integrator GIS noise wall layer.

---

[27] The VDOT noise program's website is available at www.virginiadot.org/projects/pr-noise-walls-about.asp.

[28] VDOT's contract with Fugro Roadware, which has multi-year renewable potential, includes collecting images and data on more than 12,000 directional miles of interstate and primary highways and approximately 7,700 miles of asphalt secondary roads. Over the course of the contract, Fugro Roadware will collect a variety of data for the State including faulting, rutting, roughness, forward perspective right of way and pavement digital imagery. See also www.fugro.com/news/newsdetails.asp?item=489.

Figure 12. Integrator's Noise Wall Layer.[29]

This update will help VDOT resolve earlier information discrepancies and meet 23 CFR 772's reporting requirements. VDOT will also use the new layer to model maintenance needs. For instance, maintenance staff could view catalogued images to assess the extent of damage on the walls and identify appropriate mitigation steps. Updating this layer with maintenance deficiencies will support more accurate budgeting for maintenance purposes. However, VDOT is still determining the best approach for adding information on planned walls into GIS.

VDOT uses GIS to support other noise activities in addition to capturing sound wall locations and conditions. For example, VDOT supplements civil surveys with terrain models or coverages from the Integrator DEM layer—particularly at the county level—to understand what land features exist beyond surveyed highway right of way limits. By supplementing land use information and other terrain coverages beyond the right of way, the noise model can reliably predict more accurate sound level results. VDOT noted, however, that while having civil terrain data available at a very fine and detailed level helps improve modeling accuracy, this information is not always available. A majority of the GIS terrain data[30] used to supplement the noise studies are publicly available through the Virginia Geographic Information Network (VGIN), which is run by the Virginia Information Technologies Agency (VITA), a State agency that provides IT services to other State agencies and the public.[31]

It is also considering implementing GIS to log noise-related complaints received from the public and to identify when and where past complaints have occurred. VDOT would like to be able to use this information to identify and document areas with a history of noise complaints, but has found this difficult given broader privacy concerns about linking personal identification information with the complaint. Due to these privacy concerns, it is unknown whether an internal desktop GIS tool or an enterprise GIS (available either internally or externally via a website) would be the best method for dealing with these issues.

Additionally, VDOT uses desktop GIS to develop graphics for noise reports and in conjunction with TNM. To do so, sound wall design plans are imported and georeferenced into a desktop GIS where they are converted into useable TNM input files. Steps are needed to convert VDOT's project coordinate system into a recognized projected coordinate system in GIS. After they are input and processed into TNM, the calculated (predicted) results are exported back into the desktop GIS where they can be analyzed more

---

[29] This screenshot shows noise barriers along Interstate 95 near the Virginia 123/Gordon Boulevard intersection near Woodbridge, Virginia.
[30] Virginia Base Mapping Program's (VBMP) digital terrain files are available for purchase at http://www.vita.virginia.gov/isp/default.aspx?id=8412.
[31] More information on VITA is available at www.vita.virginia.gov.

quickly and efficiently than in TNM. Using GIS, the locations of potential noise barriers are then exported out of TNM and imported into GIS where the barrier essentially undergoes a "reality check" of whether a noise barrier will be effective if it is placed at that location. The potential barriers are also evaluated to ensure that the proposed wall heights will effectively abate noise and that the wall will adequately break the line of sight with the roadway and impacted receptors (i.e., so that the roadway cannot be seen from an impacted receptor). The TNM is then used to ascertain final heights and lengths for proposed noise barriers. VDOT noted that workarounds are needed to use GIS data with TNM since the model's interface does not currently utilize a GIS platform.

GIS Data

VDOT obtains GIS data from many sources, including VGIN. Through VGIN, VDOT can access several GIS layers, including information on terrain lines, aerial imagery, orthoimagery, and digital road centerlines, which are used in conjunction with VDOT's Integrator noise layers.

VDOT's noise staff members also contact local planning divisions on an as-needed basis to ask for digital or PDF versions of development plans. Information from these plans is then georeferenced into the desktop GIS for noise modeling purposes. When necessary, VDOT can also create its own files from development plans to accurately depict any development that has occurred since a project plan was created.[32]

VDOT also extracts noise-related GIS data from other databases, such as Radford University's spatial data server[33] or from Google Earth to validate and ground-truth information.

Benefits

VDOT noted that use of GIS data has made noise modeling and noise mitigation more efficient. For example, GIS has made it easier for VDOT to compile information on receptor locations, noise levels, and other factors and input this information as a package into TNM.

Additionally, use of desktop GIS increases the level of accuracy of noise analyses. It also allows for greater flexibility and efficiency in barrier design by ensuring that the barriers have adequate termini and provide proper noise attenuation.

While there are many anecdotal indications of the benefits of GIS, it has been difficult for VDOT to comprehensively assess either the benefits or challenges related to VDOT's use of GIS, as the agency is still in the early stages of updating the GIS noise barrier layer. Nevertheless, VDOT's noise section is actively progressing with its use of GIS to assist with noise analyses.

Challenges

VDOT is addressing difficulties related to declining resources and funding, especially in regards to investments that support noise wall maintenance, collecting GIS data, or expanding GIS tools. The computing capacity required to run GIS makes it particularly important to continually invest in technology infrastructure and upgrades, but the need for continual software and data updates can make it more difficult to obtain funding. There is an added challenge of using freeware and third party non-ESRI GIS add-ons. A portion of VDOT's GIS users prefer using Google Earth to make business decisions with Integrator. However, VDOT IT policy prohibits downloading of "unapproved" software and can act as a roadblock when trying to obtain necessary tools needed for production.

Another challenge VDOT has faced is identifying how to best upload noise wall information to GIS format given the fact that a constructed wall might have some variances (in terms of specific dimensions, location, etc.) in how it is represented in a project plan. The agency is currently identifying approaches for

---

[32] 23 CFR 772.11 states that new developments must be at least in the permitting stages to be considered for noise mitigation.
[33] Radford University's spatial data server is available at http://geoserve.asp.radford.edu.

more consistently collecting information on constructed walls and adding that information to the GIS noise layer.

Future Steps

In the future, VDOT would like to conduct more on-the-ground surveys to verify the information that the consultant is collecting on noise walls across the State and help districts and residents better identify noise concerns. Conducting this survey of noise walls might cost $250,000.

VDOT is also exploring the collection and use of LiDAR to determine the locations and top of wall elevations for constructed noise barriers across the State. Costs for this effort are not available at this time.

Washington Department of Transportation (WSDOT)

Background

WSDOT's noise program is organizationally located within the agency's Environmental Services Office. There are five staff members working within the Air Quality, Noise, and Energy program.[34] Two other WSDOT staff manage the Environmental Service Office's GIS data, including its noise layers. The agency used GIS to develop a noise wall layer that displays a variety of information about the 20 miles of noise barriers in the State, including existing locations, and eligible locations for Type II retrofits.

The initial source of data for this layer was a spreadsheet-based noise wall inventory that WSDOT developed to meet Federal requirements under 23 CFR 772.

> **WSDOT uses GIS for noise analysis to:**
> - Catalog and inventory information about noise walls across the State.
> - Identify when a noise variance for construction work would be required based on jurisdictional (city/county) noise ordinances.
> - Identify potentially sensitive receptors associated with project alternatives based on land use and current orthophotos.
> - Identify appropriate noise mitigation strategies/abatement efforts.

In addition to using GIS to capture information on noise walls, WSDOT uses GIS data on an ad hoc basis for a variety of other noise analyses, such as identifying where noise might cause impacts near a roadway project and to develop maps and graphics to support these analyses.

GIS Noise Tools

WSDOT's first statewide noise wall inventory was developed in May 1981. Initially, it was a book of photographs, graphics, and text describing constructed barriers across the State. Information on Type II barriers was added to this document in 1999 when WSDOT began its Type II program. In 2005, WSDOT took information available in the document and developed a new spreadsheet-based noise wall inventory to respond to the reporting requirements of 23 CFR 772 (see Figure 13). This was developed over the course of several months using data culled from the previous inventory, project plans, as well as field visits to confirm data. There were no specific costs associated with development of this inventory besides staff time.

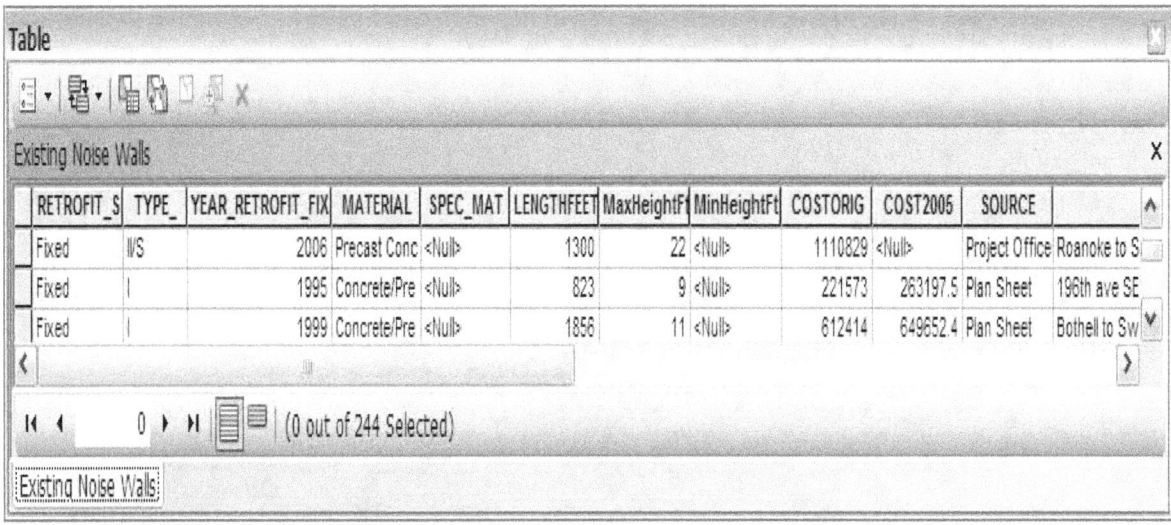

Figure 13. Screenshot of Select Fields from WSDOT's Noise Inventory (for existing walls).

---

[34] For more information on WSDOT's noise program, see www.wsdot.wa.gov/Environment/Air/Noise.htm

As a next step, WSDOT selected attributes for two datasets—proposed and existing walls—and uploaded them into the Workbench (see Figure 14), The Workbench is a GIS application customized by WSDOT that contains over 700 layers of environmental and natural resource management data, including the noise-related information. The GIS Workbench is accessible only to internal WSDOT staff, although some data are made freely available to the public through the GeoData Distribution Catalog.

Attributes added to the Workbench included the materials used for wall construction, the year the wall was built, its total cost, length, height, city or county location, source of the data, material used for construction, method of construction (e.g., precast), barrier type (Type I or Type II) and potential locations for proposed Type II noise barriers. These attributes were chosen to add to the Workbench because of the frequency with which project developers and engineers require this information. Information from WSDOT's linear referencing system[35] was also added to show where noise walls are located in relation to the State's roadways.

Figure 14. Screenshot of Existing and Proposed Noise Walls from the Workbench.

These data are updated on an as-needed basis. For example, following construction of a project that includes noise walls, WSDOT's noise staff work with project teams to obtain information about the cost, height, length, and final location (based on roadway mileposts) of noise walls that are updated into the GIS.

In the noise context, WSDOT uses the GIS Workbench primarily for project scoping and planning to identify appropriate noise abatement strategies and address noise-related queries from the public, such as whether a specific community is eligible for a Type II retrofit barrier. Workbench noise data also help support other divisions' functions. For example, WSDOT's biology office might use the data to identify the effects of noise walls on wetlands, or the hydrology office might use the data to understand how noise walls impact drainage patterns. The noise wall GIS data are also posted on a public website, WSDOT's online GeoData Distribution Catalog, to support the broader goal of sharing information both within and outside of WSDOT.[36]

TransMapper, short for "Transportation Mapper," contains a subset of information found in the Workbench.[37] TransMapper is an internal tool developed using cost-free ArcGIS Explorer and web data services. The tool contains priority data layers and base maps stored within the GIS Workbench to support numerous agency business functions (see Figure 15).

---

[35] Linear referencing systems measure the relative distance of features along a linear pathway, such as a roadway.
[36] WSDOT's GeoData Distribution Catalog is available at www.wsdot.wa.gov/mapsdata/geodatacatalog/.
[37] More information on the Workbench is available at www.wsdot.wa.gov/Environment/GIS/workbench.htm. Additional information on TransMapper is available at www.wsdot.wa.gov/NR/rdonlyres/EEDA44F7-3B51-4FF3-84E1-5830C705EAC7/0/GeographicServicesFolioJanuary2009.pdf.

Figure 15. Screenshot of TransMapper with Noise Wall Layers (subset of noise barrier layers shown in Figure 12). Green Indicates Proposed Walls and Red Indicates Built Walls.

WSDOT uses TransMapper in a variety of noise analysis activities. Staff members use the application to identify potentially sensitive receptors within project alternatives, eliminating or reducing the need to visit every site. To accomplish this, alignment lines from project alternatives, obtained as a shapefile from WSDOT's planning department, are imported into TransMapper. Using "worst-case" traffic data, staff use TNM to identify specific distances from roadway alternatives that achieve a 10-decibel level increase or produce modeled sound levels at or above 65.5 decibels (see Figure 16). Using TransMapper, staff members can then create an "area of influence" buffer around each project alternative. If potentially sensitive receptors are located within this buffer, they will be modeled in TNM to determine reasonable and feasible noise abatement.

Figure 16. Screenshot of Noise Analysis in TransMapper. (Red Lines Represent Existing Roadways. Yellow Lines Represent Buffers Associated with Different Project Alternatives. Yellow and Green Dots Represent Sensitive Receptors such as Residences).

WSDOT does not use GIS data with TNM. However, the agency does use GIS data, maps, orthoimagery, and aerial imagery to refine the scope of TNM modeling by identifying residences located next to roadways that might be impacted by traffic noise. These data are primarily obtained from the Workbench or TransMapper applications.

Potential Area of Noise Effects GIS Model

Currently, project managers and engineers do not have good tools to assess a project's early phases and whether a noise wall will be necessary. To address the need for early scoping tools, WSDOT is developing a "Potential Area of Noise Effects" GIS model using Workbench data such as annual average daily traffic, posted speed limits, volume, and vehicle class. This model will calculate the distance to the 65.5 decibel level limit[38] using emission equations provided in 23 CFR 772. The calculation result will be fed back to GIS to generate a single GIS data layer with buffers that vary in distance from the roadway according to the likelihood that noise mitigation will be necessary (Figure 17). Essentially, the model will help flag areas that are likely to need abatement structures to allow for better estimates of wall design and construction costs. The model results are currently only intended to provide a rough estimate of those needs. The model will also provide a mechanism to share and communicate information to the public.

Jurisdictions could also use information from the model to plan and zone in ways that avoid placing more land uses within or near these thresholds. In order to increase the accuracy of this tool, WSDOT intends to add land-use information to the Workbench, so that it can be incorporated into the model in the future. The model might also incorporate information on sound levels generated from TNM as well as geospatial coordinates for each noise receiver.

Figure 17. Screenshot of Output from WSDOT's "Potential Area of Noise Effects" GIS model. Different Line Widths Indicate Different Traffic Speeds and Volumes.

Benefits

WSDOT has not developed any formal metrics to assess its use of GIS. However, statistics are maintained on how many staff access specific GIS layers from the Workbench. In the future, WSDOT intends to use these statistics to evaluate which divisions or groups within WSDOT are using GIS noise layers. This could help show the extent to which staff throughout the agency rely on GIS to accomplish business tasks and help noise staff tailor the screening-level tool to best meet user needs.

In making information broadly available throughout WSDOT, the Workbench and TransMapper applications have helped staff more effectively communicate with each other and with the public. For example, prior to adding noise layers to the Workbench, WSDOT's project scoping staff would have needed to approach noise staff to ask whether a noise barrier existed in a particular location. Noise staff

---

[38] 66 decibels or greater is FHWA's criterion for what constitutes an "impacted location."

would have manually scanned project plans and maps to respond. Now, scoping staff can access these data on demand. The potential need for Type II barriers can be considered at the beginning of a project rather than at its conclusion. Additionally, the accessibility of the Workbench and TransMapper applications assists staff members with responding to public queries about noise and allows a better prediction of the need for noise variances for construction. Cataloging noise barriers in GIS also helped WSDOT create a lasting library of information about the agency's noise abatement activities, mitigating the effects of staff turnover or transition.

Furthermore, GIS has streamlined agency processes. For example, the Workbench makes it easy for staff to collect information on noise walls without having to drive to specific locations.

Future Steps

The agency is still evolving in its use of GIS to support noise analysis. Staff members intend to think more about how TransMapper and the GIS Workbench could be used to anticipate potential needs for noise analysis as well as communicating with the public and other parts of the agency. Additionally, WSDOT staff members would like to work on improving data accuracy/consistency and integrating noise wall decision documentation with GIS to improve access to decision records. Overall, WSDOT noted that GIS has provided many benefits and that "the more we learn about GIS, the more we ask ourselves how we can expand its use."

WSDOT intends to continue refining the GIS noise data layers available in Workbench and TransMapper, particularly in terms of making these data more accurate. In the next six months to one year, WSDOT anticipates developing an internal website to assist staff with cataloging and accessing records on past noise studies to help answer questions about why a previous noise decision was made. These report locations would also be available in GIS so that while scoping a proposed project, a user could open the report(s) for nearby previous studies. WSDOT will also be considering other ways to improve complaint response processing thorough better use of available information management technologies, including GIS.

Lessons Learned

WSDOT suggested the following lessons learned:

- **Offset GIS data to ensure accurate analysis.** Some noise wall GIS data points are not properly offset on the accumulated route mileage format, which can cause the wall to appear as if it is located in the centerline of a roadway when used in ArcGIS Explorer-based applications (such as TransMapper). This is because ArcGIS Explorer is not yet able to convert mileposts into offsets along the roadway. Offsetting GIS data points to the left or right of a roadway before importing data into ArcGIS Explorer applications can support more accurate analysis of noise barrier locations. WSDOT has found that it is most efficient to offset GIS locations as analysis proceeds, rather than retroactively.

- **Build comfort and expertise with GIS.** Some staff members might not be familiar with geospatial technologies. WSDOT has found that giving its personnel better access to GIS data can help build comfort levels. Ultimately, increasing comfort levels can support WSDOT's agency-wide effort to promote increased use of GIS.

- **Balance the appropriate amount: between "too much" and "just enough" detail.** WSDOT believed it was important to determine the right level of detail to include in a GIS-based tool. It is important to have enough detail that noise data can be added to models and result in useful outputs. However, if a GIS tool is used primarily during scoping and early project development stages, including a high level of detail could be difficult or unnecessary. Furthermore, noise can be very specific to sites, conditions, and times of day, so collecting detailed noise data can result in large packages of information that are unique to a single project. Some details might not be applicable to other projects and thus would not be a high priority for adding to a GIS tool.

APPENDIX A: LIST OF INTERVIEW PARTICIPANTS

Agency	Name*	Title	Phone	Email
FHWA	**Adam Alexander**		202-366-1473	adam.alexander@dot.gov
	Mark Ferroni		202-366-3233	mark.ferroni@dot.gov
	Mark Sarmiento		202-366-4828	mark.sarmiento@dot.gov
Volpe Center	**Jaimye Bartak**	Transportation Analyst	617-494-2451	jaimye.bartak.ctr@dot.gov
	Alisa Fine	Community Planner	617-494-2310	alisa.fine@dot.gov
Caltrans	**Jim Andrews**	Senior Transportation Engineer	916-653-9554	jim.andrews@dot.ca.gov
	Lefteris Koumis	Senior Transportation Engineer	916-653-0053	lefteris_koumis@dot.ca.gov
Florida DOT	**Mariano Berrios**	Environmental Programs Administrator	850-414-5250	mariano.berrios@dot.state.fl.us
	Pete McGilvray	Environmental Quality Performance Administrator	850-414-5330	peter.mcgilvray@dot.state.fl.us
	Matt Muller	Technology Resource Manager	850-414-5329	matthew.muller@dot.state.fl.us
Maryland SHA	**Ken Polcak**	Noise Abatement Design & Analysis Team Leader	410-545-8601	kpolcak@sha.state.md.us
	Mike Sheffer	Assistant Division Chief / GIS Coordinator	410-545-5537	msheffer@sha.state.md.us
North Carolina DOT	**Joe Rauseo**	Acoustic Engineer	919-707-6084	jarauseo@ncdot.gov
Ohio DOT	**Noel Alcala**	Noise and Air Quality Coordinator	614-466-5222	Noel.alcala@dot.state.oh.us
	Craig Ayers	Noise and Air Quality Intern	614-995-2187	craig.ayers@dot.state.oh.us
	Gary Penn	GIS Coordinator	614-466-7100	gary.penn@dot.state.oh.us
Tennessee DOT	**Jim Ozment**	Manager, Social and Cultural Resources Department	615-741-5373	jim.ozment@tn.gov
	Ann Epperson	Transportation Manager 1	615-253-2470	ann.epperson@tn.gov
Virginia DOT	Geraldine Jones	CEDAR Administrator / Environmental GIS Coordinator	804-786-6678	Geraldine.Jones@vdot.virginia.gov
	Tiffany Abbondanza		804-371-4954	tiffany.abbondanza@vdot.virginia.gov

Agency	Name*	Title	Phone	Email
	Paul Kohler	Air Quality, Noise & Energy Section Manager	804-371-6766	Paul.Kohler@vdot.virginia.gov
	Josh Kozlowski	Noise Abatement Specialist	804-371-6829	Joshua.Kozlowski@vdot.virginia.gov
Washington DOT	**Larry Magnoni**	Air and Acoustics Specialist	206-440-4544	Larry.Magnoni@wsdot.wa.gov
	Tim Sexton	Air/Acoustics/Energy Policy Manager	206-440-4549	Timothy.Sexton@wsdot.wa.gov
	Jim Laughlin	Air/Acoustics/Energy Technical Manager	206-440-4643	Jim.Laughlin@wsdot.wa.gov
	Elizabeth Lanzer	Program Manager, Environmental Information	360-705-7476	Elizabeth.Lanzer@wsdot.wa.gov
	Kathy Prosser	GIS/GPS Specialist	360-705-7498	Kathy.Prosser@wsdot.wa.gov

*Bolded name indicates participation in both interviews and the peer exchange.

Appendix B: Peer Exchange Agenda

Goal*: Share lessons learned, best practices, and challenges in using GIS to support noise modeling and analysis.*

Monday, April 23

1:00 – 1:30   **Welcome, Introductions, and Background** - *FHWA (Adam Alexander and Mark Sarmiento) and Tennessee DOT*

1:30 – 2:30   **Overview of FHWA GIS and Noise Activities** - *FHWA*

Break

2:45 – 3:45   **Demonstrations/Presentations 1**
- Tennessee DOT – Type II Inventory/Spreadsheet
- Caltrans – Sound Wall Inventory

3:45 – 4:45   **Roundtable 1: Reporting** - *All Participants*
- What are the potential uses and purposes of a sound wall inventory (including and beyond meeting 23 CFR 772 requirements)?
- How does GIS support developing sound wall inventories? What are the benefits and challenges?
- What types of data and maintenance are involved with developing a sound wall inventory?

4:45 – 5:00   **Day 1 Key Points/Wrap-Up** - *FHWA*

6:00   **Informal Dinner** (Big River Restaurant - 111 Broadway, Nashville, TN 37201)

Tuesday, April 24

8:00 – 8:15   **Day 1 Re-cap** - *FHWA*

8:15 – 9:15   **Demonstrations/Presentations 2**
- Florida DOT – Efficient Transportation Decision-Making Process (ETDM) and Environmental Screening Tool (EST)
- North Carolina DOT – Use of GIS for Decibel Tolerance Levels

9:15 – 10:15   **Roundtable 2: Data** - *All Participants*
- What types of data have helped support successful noise assessments/activities?
- How does/can GIS play a role in helping gather, collect, and/or maintain data?
- What are best practices for using GIS for these purposes?
- How can Google/Bing (or other mapping applications) support noise assessments?

Break

10:30 – 11:15   **Roundtable 3: Applying GIS/Geospatial Data to Assess Noise**
- How can GIS be better integrated into TNM/modeling? Should it be?
- What are other uses for GIS or geospatial data in assessing or mitigating noise (e.g., project-level screening, identifying contours, wildlife impacts)?

- Can States share any examples? What are the benefits, challenges, and lessons learned?

11:15 – 11:45 **Demonstrations/Presentations 3**
- Ohio DOT- Noise Barrier Inventory/Use of GIS for Traffic Noise Model

Lunch

12:45 – 1:45 **Demonstrations/Presentations 3 (continued)**
- Maryland SHA – Noise Complaint Inventory
- Washington DOT – Planned GIS Screening Layer/TransMapper Tool

1:45 – 2:30 **Roundtable 4: Assessing Potential for GIS in Noise Assessments –** *All Participants*
- How does/could GIS help States respond to current (or new) FHWA noise requirements?
- What are the benefits and challenges of GIS in doing so?
- What funding/training or other resources are needed to make the most of GIS in noise applications?
- How can staff effectively translate the benefits of GIS to agency leadership?

2:30 – 2:45 **Day 2 Key Points/Wrap-Up -** *FHWA*

2:45 **Adjourn**

APPENDIX C: PEER EXCHANGE ROUNDTABLE QUESTIONS

Roundtable 1: Reporting
- How are States using GIS to respond to the reporting requirements of 23 CFR 772?
- How does GIS support developing sound wall inventories? What are the benefits, challenges, and lessons learned?
    - What are some examples of GIS platforms/tools for sound wall inventories? How are these tools used? What are their outputs?
    - Is GIS useful for keeping track of sound wall conditions?
    - What are best practices in designing a GIS sound wall inventory?
    - Does the time/cost investment match the benefits of having a GIS-based sound wall inventory?
- Do sound wall inventories contribute to use of TNM? How?
- How does developing a sound wall inventory support other noise-related activities?
- What types of data and maintenance activities are involved with developing GIS-based sound wall inventories?
    - What data can be obtained from existing sources? What needs to be collected in the field? Are there common data "gaps"?
    - What are the challenges related to collecting/gathering data for inventories?
    - How often should the sound wall inventory be updated to be the most useful?

Roundtable 2: Data
- What types of data have helped support successful noise assessments/activities?
    - What types of data have been the most useful for planning, accurate noise modeling, and abatement?
    - What are some innovative examples?
- What are best practices and challenges in collecting, gathering, or maintaining GIS data for noise purposes?
    - How can GIS assist in sharing data? Is there a demand for access to noise data from the public and/or other State agencies?
    - Is it difficult to collect data using in-house resources? When is it helpful to have consultant assistance?
- How can Google/Bing (or other mapping applications) support noise assessments?
    - What features do these mapping applications offer that are useful in noise analysis?
    - What would make these mapping applications more useful for noise analysis?
    - What are the benefits, challenges, and lessons learned in using mapping applications?

Roundtable 3: Other Applications of GIS/Geospatial Data to Assess Noise
- How can GIS improve TNM/noise modeling? Is there a need for GIS?
- What are other uses for GIS or geospatial data in assessing or mitigating noise (e.g., project-level screening, identifying contours, wildlife impacts, noise abatement)?
    - Is there currently capacity within agencies to conduct these analyses?
    - What topics are the most important to pursue?
- Can States share any examples? What are the benefits, challenges, and lessons learned?
- How can States use GIS to support communication with the public about noise? Are there any current examples, including web-based mechanisms?

Roundtable 4: Assessing Potential for GIS in Noise Assessments
- What is your view of the short- and long-term potential of GIS to improve noise analysis?
- What are the benefits and challenges of GIS in responding to the current (or new) FHWA noise requirements? How does/could FHWA help States use GIS for this purpose?
- What resources (e.g., funding, training) are needed to enhance the effectiveness of GIS in noise analysis?

- How can State DOT staff effectively translate the benefits of GIS to agency leadership?
    - What efficiencies in cost/time does GIS produce? How can State DOTs measure and communicate these efficiencies?
    - Are there ways to more formally evaluate uses of GIS to support noise analysis? Are there any examples from State DOTs?

APPENDIX D: ADDITIONAL RESOURCES

This appendix includes examples of resources that support GIS tools for noise. FHWA does not endorse any specific resource.

- AASHTO GIS-T Task Force Annual Symposium shares knowledge and experience among state transportation agencies in using geospatial data and technologies.

- Transportation Research Board Committee on Geographic Information Science and Applications (ABJ60) determines and develops research need statements.

- National Highway Institute's "Applying Spatial Data and GIS to Transportation" Course # 151039 is a beginner-level course that targets agency executives to provide a basic understanding of what GIS is and what it can accomplish.

- FHWA Workshops on GIS are provided through the FHWA Resource Center.

- FHWA GIS outreach:
    o State GIS efforts and applications database;
    o "GIS in Transportation" Newsletter, Webcast and Website;
    o GIS case studies and peer exchanges.

APPENDIX E: SELECT STATE DOT APPLICATIONS OF GIS FOR NOISE

State	Developed GIS/Geospatial Noise Barrier Inventory	Use GIS for Project Screening, Planning & Programming for Noise Issues	Use GIS Data/Tools in TNM Modeling	Developed or Uses Enterprise or Web-Based GIS Tool	Use Web-Based GIS Mapping Service (i.e. Google, Bing)	Log Noise Complaints Geospatially	Use GIS for Noise-Related Communication with the Public
Caltrans	Yes (includes video)	No	No	Yes (developed noise barrier mapping tool using ESRI FlexViewer)	Yes (for reference purposes, and for integration into noise wall inventory)	No	Yes (noise barrier tool publicly accessible)
FDOT	Yes (retrofitting existing inventory to respond to 23 CFR 772)	Yes (noise issues are considered as an aesthetic issue in a statewide web-based GIS tool in the planning/programming phases for all major transportation projects)	No	Yes (FDOT Environmental Screening Tool)	No	No	Yes (ETS publicly accessible)
MDSHA	Yes (under development)	No	No	Yes (developing noise barrier tool for MD iMap)	Yes (for reference purposes)	No	Yes (MD iMap publicly accessible)
NCDOT	No	Yes (uses GIS data to complete TNM noise models for all project alternatives)	Yes (uses GIS extensively for input into TNM model)	Yes (uses NC One Map, a statewide GIS repository, to obtain GIS data)	No	No	No
ODOT	Yes (georeferenced walls in spreadsheets)	No	No	No	Yes (uses Google to obtain noise barrier coordinates)	No	Yes (to locate noise complaints)
TDOT	Yes (under development)	No	Yes (uses GIS data on a project-by-project basis)	Yes (TNM, but noise barrier data not yet included)	Yes (for reference purposes)	No	No (plans to do so in the future)
VDOT	Yes (currently updating old database)	No	Yes	Yes (the noise barrier layer is available in Integrator)	Yes (for reference purposes)	No (under consideration)	No

WSDOT	Yes (includes proposed and existing noise barriers)	Yes (developing a GIS-based model that uses traffic data along with noise emission curves to identify rudimentary noise impact areas for preliminary screening purposes)	No	Yes (WSDOT's Workbench includes all agency GIS data, including some noise-related data)	Yes (for reference purposes)	No	Yes (existing and proposed noise barriers available online)

APPENDIX F: REQUIRED NOISE BARRIER INVENTORY DATA

As outlined in 23 CFR 772, FHWA requires State DOTs to submit (on a triennial basis) the following information on noise barriers:
- Type of Abatement
- Cost
    - Overall cost
    - Unit cost per/sq. ft.
- Average Height
- Length
- Area
- Location
    - State
    - County
    - City
    - Route
- Year of Construction
- Average Insertion Loss/Noise Reduction (as reported by the model in the noise analysis)
- Noise abatement category(ies) protected
- Material(s) Used
    - Precast concrete
    - Berm
    - Block
    - Cast-in-place concrete
    - Brick
    - Metal
    - Wood
    - Fiberglass
    - Combination
    - Plastic (transparent, opaque, other)
- Features
    - Absorptive
    - Reflective
    - Surface texture
- Foundation
    - Ground mounted
    - On structure
- Project Type
    - Type I
    - Type II
    - Optional project types (e.g., State funded, county funded, tollway/turnpike funded, other, unknown)

www.ingramcontent.com/pod-product-compliance
Lightning Source LLC
Chambersburg PA
CBHW081859170526

45167CB00007B/3081